Make Your Own PCBs with EAGLE™

Make Your Own PCBs with EAGLE™

From Schematic Designs to Finished Boards

Simon Monk

New York Chicago San Francisco
Athens London Madrid
Mexico City Milan New Delhi
Singapore Sydney Toronto

McGraw-Hill Education books are available at special quantity discounts to use as premiums and sales promotions or for use in corporate training programs. To contact a representative, please visit the Contact Us page at www.mhprofessional.com.

Make Your Own PCBs with EAGLE™:
From Schematic Designs to Finished Boards

1 2 3 4 5 6 7 8 9 0 DOC/DOC 1 2 0 9 8 7 6 5 4

ISBN 978-0-07-181925-1
MHID 0-07-181925-8

This book is printed on acid-free paper.

Sponsoring Editor
Roger Stewart

Editing Supervisor
Stephen M. Smith

Production Supervisor
Pamela A. Pelton

Acquisitions Coordinator
Amy Stonebraker

Project Manager
Patricia Wallenburg, TypeWriting

Copy Editor
James K. Madru

Proofreader
Alison Shurtz

Indexer
Claire Splan

Art Director, Cover
Jeff Weeks

Composition
TypeWriting

To Stephen,

from a very proud Dad.

About the Author

Dr. Simon Monk (Preston, UK) has a degree in cybernetics and computer science and a Ph.D. in software engineering. He spent several years as an academic before he returned to industry, co-founding the mobile software company Momote Ltd. He has been an active electronics hobbyist since his early teens and is a full-time writer on hobby electronics and open-source hardware. Dr. Monk is the author of numerous electronics books, specializing in open-source hardware platforms, especially Arduino and Raspberry Pi. He is also co-author with Paul Scherz of *Practical Electronics for Inventors*, Third Edition.

You can follow him on Twitter, where he is @simonmonk2.

Contents

Acknowledgments

Many thanks to all those at McGraw-Hill who have done such a great job in producing this book. In particular, thanks to my editor Roger Stewart and to Patty Wallenburg.

I am most grateful to Mike Basset for his technical review of the material.

And, last but not least, thanks once again to Linda, for her patience and generosity in giving me space to do this.

Make Your Own PCBs with EAGLE™

CHAPTER 1
Introduction

In this chapter, you will learn how to install EAGLE™ Light Edition and will discover the various views and screens that make up an EAGLE project. EAGLE (Easily Applicable Graphical Layout Editor) is a product of the German company Cadsoft. The company is now a subsidiary of Premier Farnell, which also owns Newark Electronics in the United States and CPC in the United Kingdom.

The software has been around for many years, and despite having a user interface that can seem a little daunting to newcomers, it is a powerful and flexible product. It has become a standard for hobby use primarily because of its freeware version and the large set of component libraries and general adoption as the standard tool for open-source hardware (OSH) providers such as Sparkfun and Adafruit. Generally, you will find EAGLE design files available for their OSH products as well as for high-profile products such as the Arduino family of circuit boards.

Printed Circuit Boards

Because you are reading this book, you probably want to make a printed circuit board (PCB) and already have a basic understanding of what exactly a PCB is and how it works. However, PCBs come with their own set of jargon, and it is worth establishing exactly what we mean by *vias*, *tracks*, *pads*, and *layers*.

The main focus of the book will be on making double-sided professional-quality circuit boards. This book assumes that you will design circuit boards and then e-mail the design files to a low-cost PCB fabrication service (as low as US$10 for 10 boards) that will actually make the boards. The making of PCBs at home is now largely redundant because they can generally be made at lower cost and to a better standard than home PCB etching, with all its attendant problems of handling and disposing of toxic chemicals or the need for expensive milling machines.

Figure 1-1 shows the anatomy of a two-layer PCB. You will see exactly how this PCB was designed later in this book, where it is used as an example. For now, let's briefly explain the anatomy of a PCB. Referring to Figure 1-1, we have the following:

- *Pads* are where the components are soldered to the PCB.
- *Tracks* are the copper tracks that connect pads together.
- *Vias* are small holes through the board that link a bottom and top track together electrically. Tracks on the same layer cannot cross, so often, when you are laying out a PCB, you need a signal to jump from one layer to another.
- *Silk-screening* refers to any lettering that will appear on the final board. It is common to label components and the outline of where they fit so that when it comes to soldering the board together, it is easy to see where everything fits.
- *Stop mask* is a layer of insulating lacquer that covers both sides of the board except where there are pads.

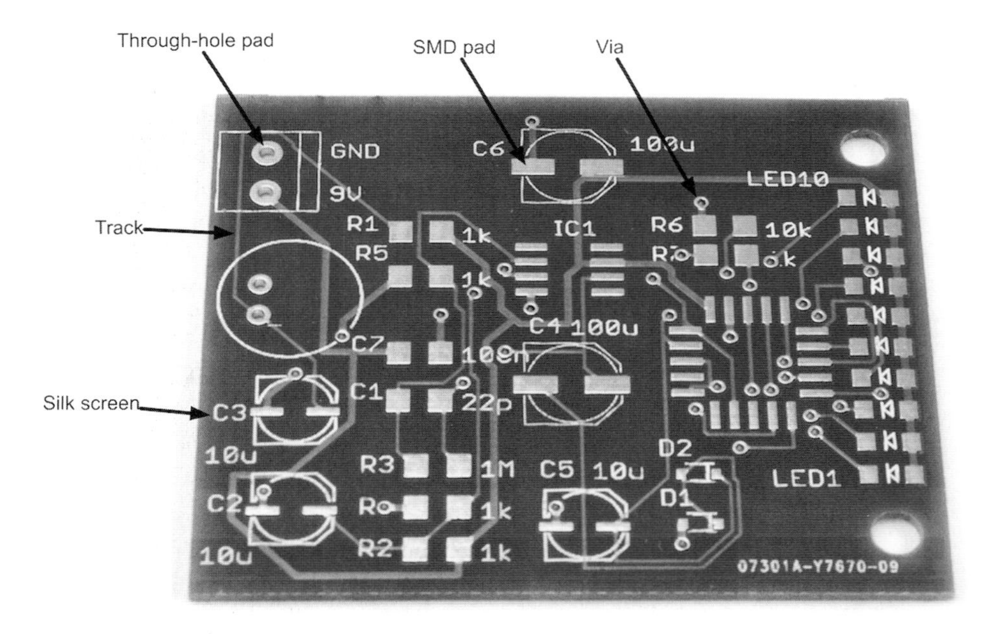

Figure 1-1 Anatomy of a double-sided PCB.

Surface Mount and Through Hole

Pads are either *though hole*, where components with leads are pushed through from the top, soldered underneath, and then the excess lead snipped off, or *surface mount*, where the components are soldered to the top of the pad. Figure 1-2 shows a board that contains both surface-mount and through-hole components.

Surface-mount components are often referred to as *surface-mount devices* (SMDs) and are replacing through-hole components in most commercial products. This is so because SMD components are smaller and cheaper than their through-hole counterparts, and the boards that use them are also easier to make. You will also see the term *surface-mount technology* (SMT) used.

In commercial surface-mount PCB production, and increasingly for hobbyists, boards are soldered by creating a mask that allows solder paste to be deposited on the pads, then the components are placed precisely on the pads, and then the whole board is baked in an oven that melts the solder paste, soldering the components without the difficulty of soldering each component separately.

SMD ovens are still too expensive for most hobbyists, but many people have had success modifying toaster ovens to operate at the high and precisely controlled temperatures required. Such experiments usually require the safety features of the

FIGURE 1-2 Through-hole and surface-mount components.

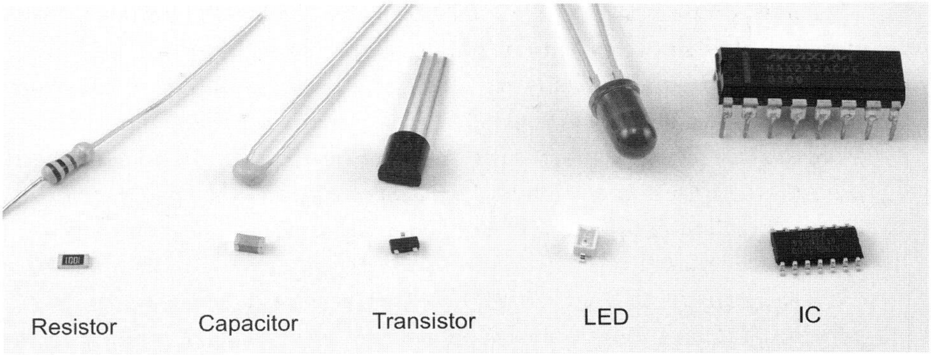

FIGURE 1-3 A selection of through-hole and surface-mount devices.

toaster oven to be disabled and are therefore often referred to as "fire starters" for good reason. However, like so many things in life, with care, common sense, and a watchful eye, such things can be made to work safely.

The choice of surface-mount versus through-hole design is less cut and dried for the hobbyist just wanting to make one or two boards for a specific project. For a single project that is never intended to be made as a commercial product, through-hole design is much simpler to solder by hand. Through-hole component leads are nearly always at least 0.1 in. apart, whereas surface-mount chips can have pins that are just 0.5 mm apart. Although many SMDs are easy enough to solder by hand, many others are just too small.

Figure 1-3 shows a selection of electronic components in both surface-mount and through-hole "flavors." As you can see, the SMDs are very much smaller than their through-hole equivalents. This generally means that you can get a lot more of them on the same area of a PCB.

Prototyping

Ultimately, if you want to produce something of professional quality, then PCBs are the only way to go. However, while you are prototyping a design, it is a very good idea to test out your design before you start getting PCBs manufactured. Every time you find something wrong with your design and have to get a new batch of PCBs made, you will be increasing costs, both in time and in money. It is far better to get the design as perfect as possible before you commit to a board. This is a bit like writing a book—you wouldn't print and bind the first draft; you need to be certain that the book is how you want it before you commit to paper.

This is a book about the EAGLE PCB and building PCBs. It is not an electronics primer, so if you need to learn more about electronics in general, then take a look at the books *Hacking Electronics* and *Practical Electronics for Inventors*, both from TAB Books.

Assuming that you have a schematic diagram for what you want to build, there are a number of useful construction techniques that you can use to build your prototypes quickly and easily.

Solderless Breadboard

Solderless breadboard (Figure 1-4) is very useful for quickly trying out designs before you commit them to solder. You poke the leads of components into the sockets, and metal clips behind the holes connect all the holes on a row together.

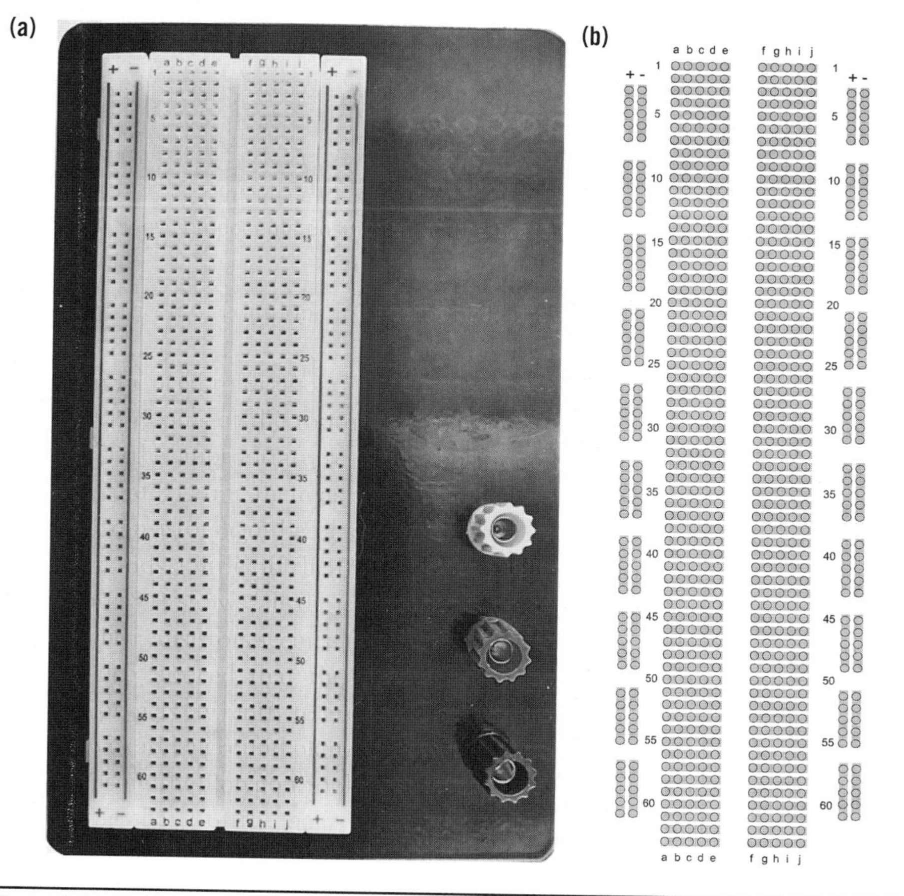

FIGURE 1-4 Solderless breadboard.

Breadboard comes in all shapes and sizes, but a big one is probably most useful. The breadboard in Figure 1-4 has 63 rows by 2 columns with two supply strips down each side. It is also mounted on an aluminum base with rubber feet to stop it from moving about on the table. This is a very common size of breadboard, and most suppliers will have something similar.

Figure 1-4*b* shows how the conductive strips are arranged underneath the plastic top surface of the board. All the holes that share a common gray area beneath are connected together in rows of five connectors. The long strips down each side are used for the power supply to the components, one positive and one negative. They are color-coded red and green.

Breadboards are often modular and will clip together in sections to make as big a board as you need. Figure 1-5 shows an example of a simple breadboard prototype.

The main advantage of a solderless breadboard is that it's, well, solderless. Thus you can quickly and easily change the design just by unplugging components and leads as you need to. The disadvantage is that wires can fall out and leads of components can touch, so a breadboard is only good for the first pass of a prototype. It probably would not be wise to deploy a breadboard-built design for real use. Eventually, something would work loose, and the prototype will stop

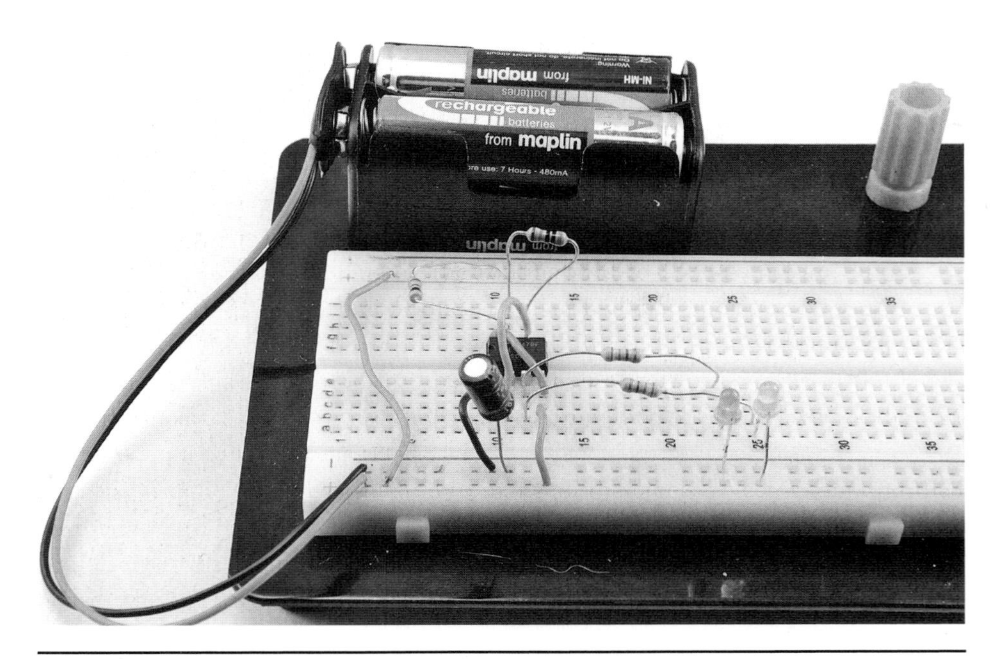

FIGURE 1-5 A breadboard prototype.

working. For something more durable as a prototype, there is really no substitute for soldering.

The other disadvantage of breadboard is that the layout is fixed, so components end up very spaced out from each other, often with a large number of jumper wires linking everything together.

Perfboard

Perfboard (perforated board) is one of a number of types of board designed specifically for prototyping. It is made from the same material as a PCB but has no copper on it. It is just a board with an array of holes in it on a 0.1-in. grid (Figure 1-6).

Component leads are pushed through from the top and soldered together underneath using either their leads (if they will reach) or lengths of solid-core wire. The perfboard effectively provides a rigid structure to keep the components in position.

A variation on perfboard called *protoboard* is just like perfboard except that behind each hole is a copper pad. The pads are not connected together, but they

FIGURE 1-6 A perfboard prototype.

serve to hold the components tight to the board. This arrangement does, however, make it more difficult to move a component once it is soldered. Generally, if a design uses dual in-line (DIL) integrated circuits (ICs), then protoboard with solder pads is easier to use than regular perfboard.

The advantage of perfboard and protoboard is that the layout of the components can be closer to the schematic diagram because you are not constrained to using fixed strips of connectors. Such designs can be strong enough to deploy in a project permanently.

Stripboard

Stripboard (Figure 1-7) is a bit like general-purpose PCB. It is a perforated board with conductive strips running underneath, rather like breadboard. The board can be cut to the size you need, and components and wires are soldered onto it.

Figure 1-7 Stripboard prototype.

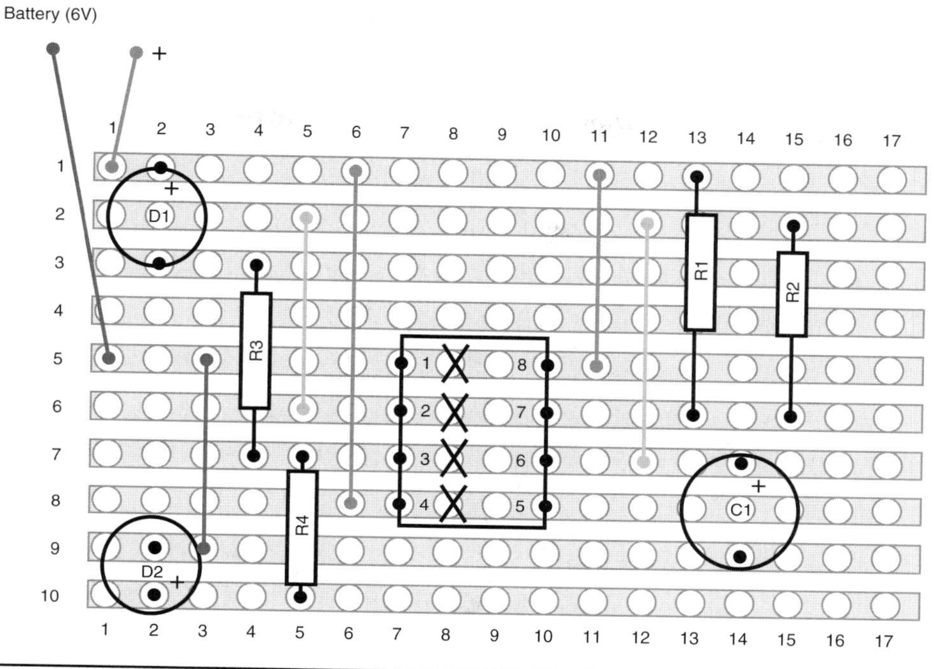

Figure 1-8 Stripboard layout.

As with breadboard, laying out a project on stripboard requires a bit of skill to rearrange the logical layout of the schematic into something that will work with the strips of the stripboard. Figure 1-8 shows the stripboard layout for the prototype in Figure 1-7.

The X's underneath the IC are breaks in the track, which are made with a drill bit, and one of the goals of a good stripboard layout is to try to avoid too many breaks having to be made in the track. Breaks are unavoidable for an IC such as this. If we did not make them, pin 1 would be connected to pin 8, pin 2 to pin 7, and so on, and nothing would work.

Installing EAGLE Light Edition

Having prototyped your design and being sure that it is time to start making PCBs, let's get on with installing and configuring EAGLE, the design software that we will use to create PCBs. One of the great things about EAGLE is that it is available for Windows, Mac, and Linux platforms. Thus the first step in installation is to go to www.cadsoftusa.com and click on the "Downloads" button. The instructions are for Version 6.3 of EAGLE.

Select the download for your platform, and then follow the instructions in the section for your operating system below.

Installation on Windows

For a Windows installation, you will need a machine running Windows XP, Vista, or 7. Download the self-extracting archive (`eagle-win-6.3.0.exe`), and allow it to run. After the file has been unzipped, the dialog shown in Figure 1-9 will open. Click "Setup" to start the installation process (Figure 1-10).

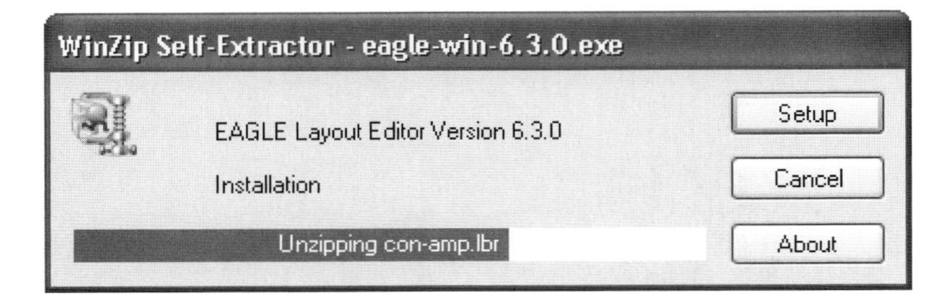

Figure 1-9 The EAGLE self-extracting archive.

Figure 1-10 The EAGLE installer.

You can just accept the defaults most of the way through the installation until you get to the "EAGLE License" step. Here you should select the option "Run as Freeware" (Figure 1-11).

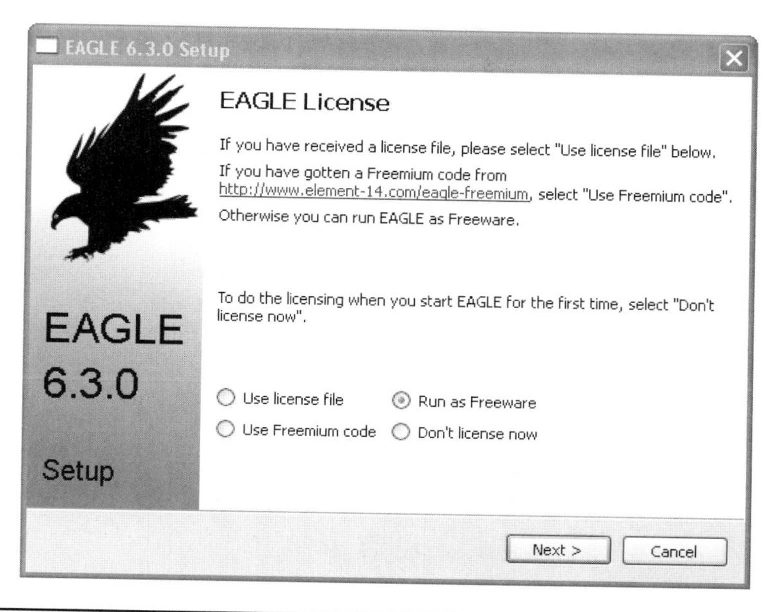

Figure 1-11 Installing EAGLE as freeware.

Eventually you will get confirmation that the installation is complete. You will find that the installer has added a new program group to the Start menu from which you can launch EAGLE. You can now skip ahead to the "First Run" section.

Installation on Mac

When installing on a Mac, EAGLE is distributed as a zipped package installer rather than a disk image. After downloading the file `eagle-mac-6.3.0.zip`, the file will extract to a package installer `eagle-6.3.0.pkg`. Double-click it to start the installation process (Figure 1-12).

You can accept the default options for the whole installation, but unlike the Windows installation, you will not be prompted to say what kind of license you are using. Instead, this will happen the first time you run the software (Figure 1-13). Select the option "Run as Freeware."

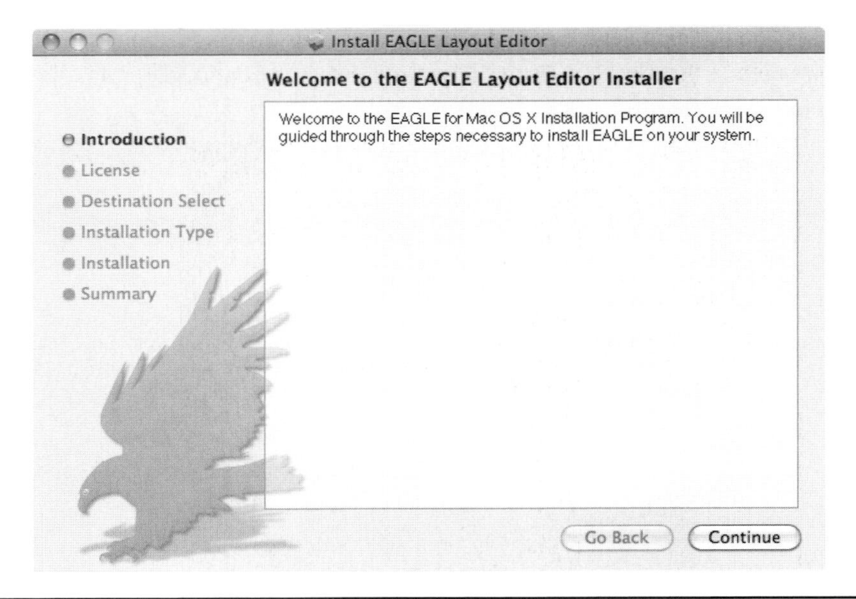

Figure 1-12 The Mac installer.

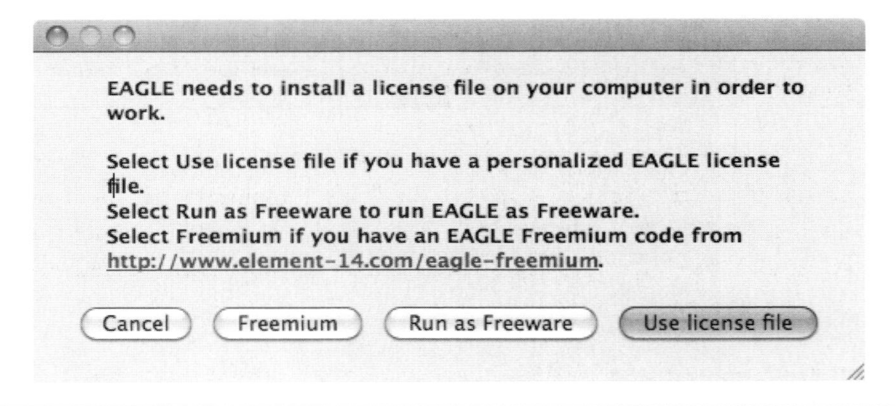

Figure 1-13 Selecting license type on first run (Mac).

Installation on Linux

Having downloaded the installation file for Linux (`eagle-lin-6.3.0.run`), try running it by opening a terminal session, changing to the "Downloads" directory, and running the following command:

```
sh ./eagle-lin-6.3.0.run
```

If you are lucky, the installer will run in a similar fashion to the Windows installer. If you are unlucky, you might see something like Figure 1-14.

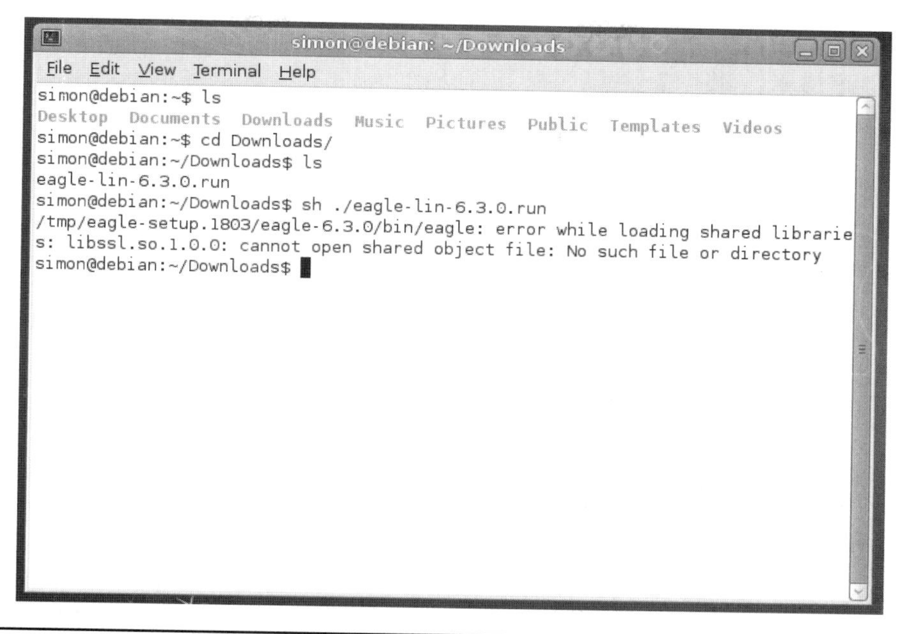

FIGURE 1-14 Installation failure on Linux.

The root cause of this is incompatible versions of two libraries (`libssl.so.1.0.0` and `libcrypto.so.1.0.0`). EAGLE requires 32-bit versions of these libraries, and resolving this problem is not always easy. Useful pointers on this can be found at http://blog.raek.se/2012/01/06/running-cadsoft-eagle-version-6-in-ubuntu-gnulinux/.

First Run

Whatever operating system you are using, launch EAGLE. The following screen shots are all for Windows, but you will find that it looks much the same in Mac and Linux.

The first time you run EAGLE, you will get something like the message shown in Figure 1-15.

This is asking your permission to create a documents folder into which EAGLE will store the files for the projects that you are working on. You should say "Yes" to this.

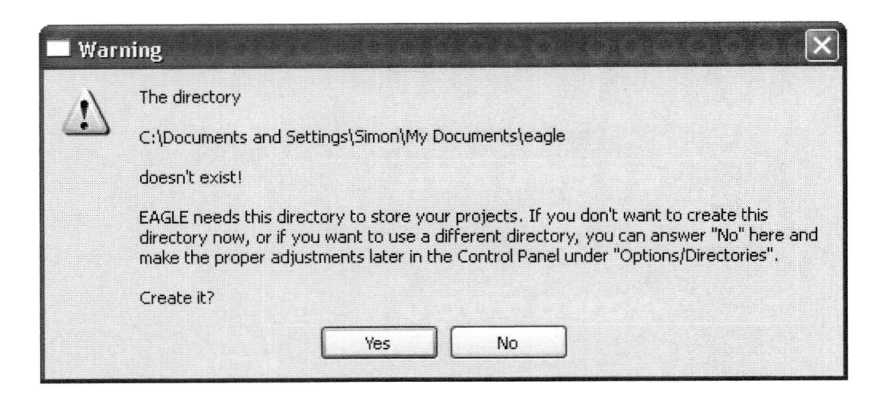

Figure 1-15 Confirming creation of a documents folder.

If all is well, EAGLE will then open up the EAGLE Control Panel (Figure 1-16).

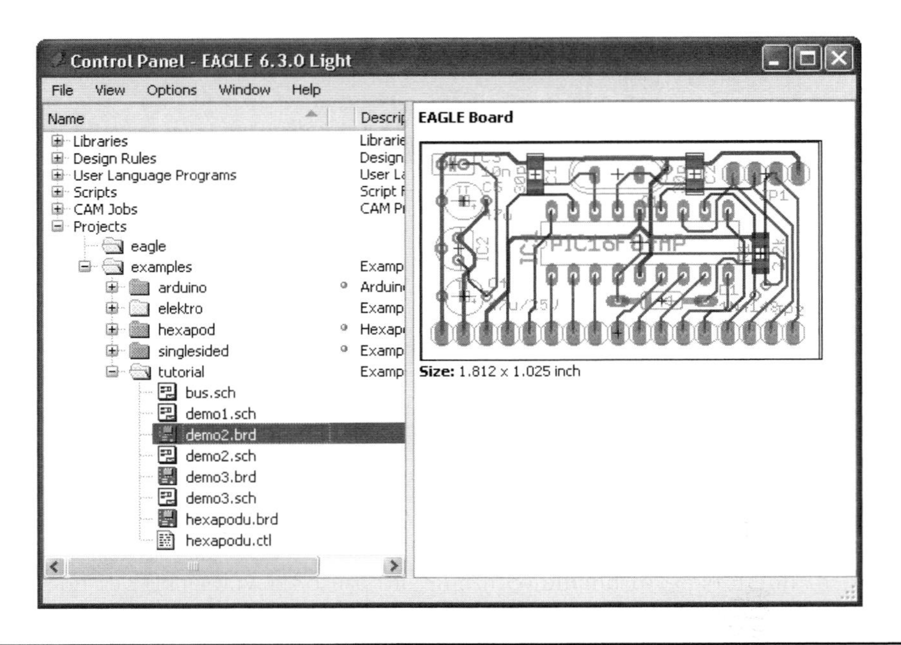

Figure 1-16 EAGLE Control Panel.

You will mostly use the Control Panel as a convenient way to access your projects. It also allows you access to all the other different documents that EAGLE uses. These includes

- Libraries of components

- Design rules that specify spacing between tracks and all manner of other things
- User-language programs (You can write your own and download programs to extend EAGLE.)
- Scripts that automate EAGLE activities
- CAM (computer-aided manufacturing) jobs that specify how the EAGLE design is converted into files suitable for PCB fabrication

Load an Example Project

Within a project, the two main files you will be using are the schematic file, which will have the extension `.sch`, and the board layout file, which will have the extension `.brd`.

Navigate down the folders in "Projects" to `demo2.brd`, as shown in Figure 1-16, and then double-click both `demo2.brd` and `demo2.sch` to open both the board layout and the schematic (Figure 1-17). Later on, when you are working on a project, you will need to make sure that the schematic and board files are open for the project, or they can become out of step, leading to all sorts of problems.

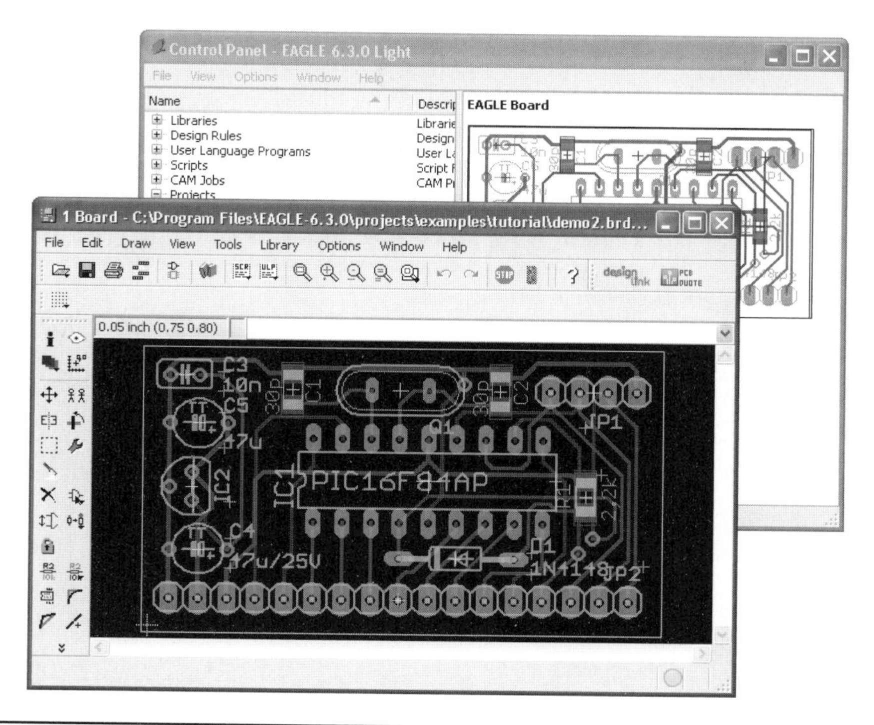

FIGURE 1-17 Example board and schematic files.

Install Third-Party Software

As part of the process of setting up our environment, we will download some useful things from the Internet that will just help to get us started and make EAGLE that much easier to use.

Installing the Adafruit and Sparkfun Libraries

You are going to load up two libraries of parts, one from Sparkfun and one from Adafruit. Although EAGLE comes with a huge collection of parts organized into libraries, the collection of parts is so huge that it can be very difficult to "see the wood for the trees" and find the part that you want. However, the Adafruit and Sparkfun libraries offer simplified lists of the most common components, and of course, they include components that the companies sell.

To download the Sparkfun library, navigate to https://github.com/sparkfun/SparkFun-Eagle-Libraries with your browser, and select the option to download a zip file (the icon looks like a cloud with an arrow pointing down).

Once the zip file has downloaded, the `.lbr` files that it contains all need to be extracted into the `.lbr` folder inside your EAGLE installation folder (Figures 1-18 and 1-19).

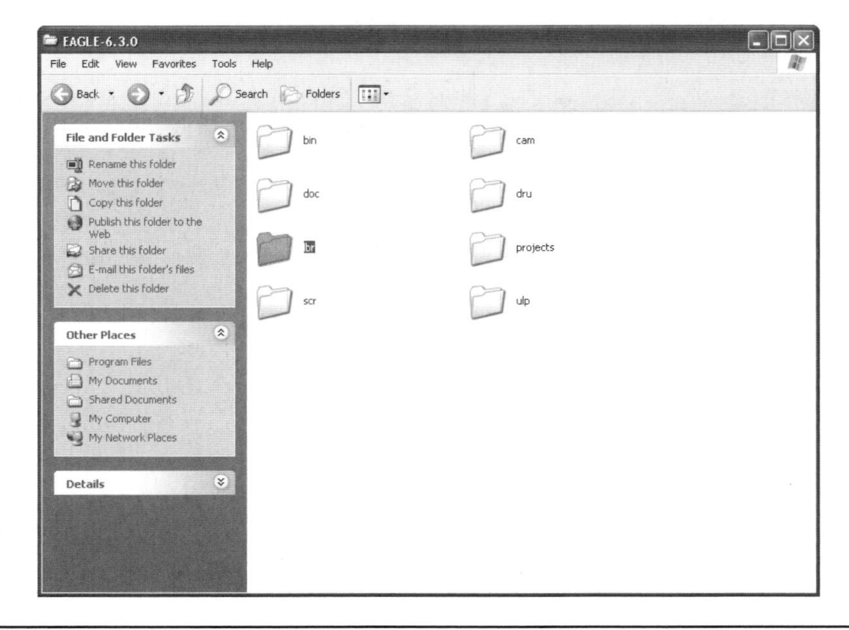

Figure 1-18 EAGLE `.lbr` folder.

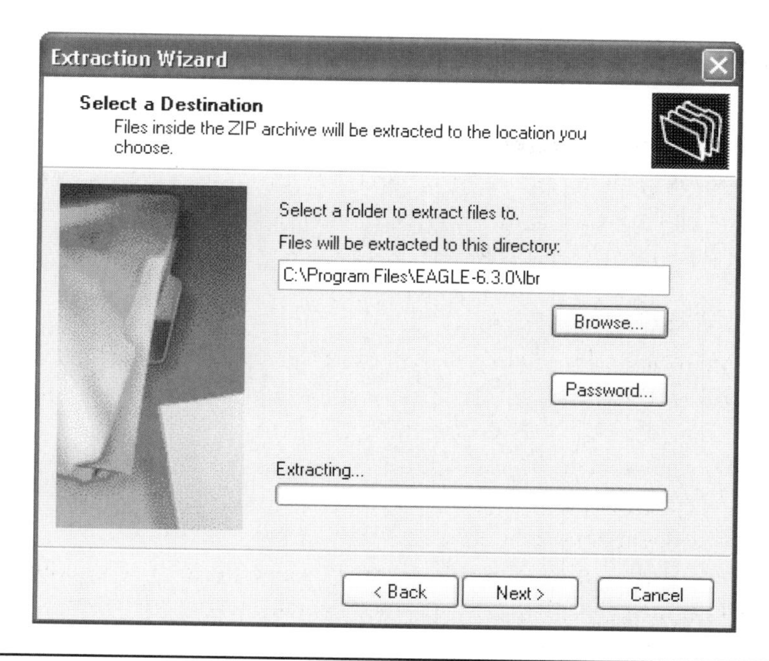

FIGURE 1-19 Extracting files straight into the EAGLE `.lbr` folder.

This will add about 17 `.lbr` files from `LilyPad-Wearables.lbr` to `SparkFun-Sensors.lbr` to the `.lbr` directory. You can remove the README file once it has been unzipped.

The installation for the Adafruit is a similar process. First navigate to http://github.com/adafruit/Adafruit-Eagle-Library and again download using the zip option and unzip the contents in the same way as you did the Sparkfun library. However, the Adafruit parts are all contained in a single library rather than a collection. Thus you can, if you wish, just add the single `.lbr` file rather than contain it in a folder.

To check that both sets of libraries have been installed, go back to your EAGLE Control Panel and expand the "Libraries" tab. You should see a list something like Figure 1-20.

Installing the Sparkfun Design Rules

Design rules specify all sorts of things about the design, the gaps between pads and tracks, and other things like that. Sparkfun has put together a useful default set of design rules that we will use in this book, so it makes sense to install them now so that you can use them in any of the EAGLE projects you create.

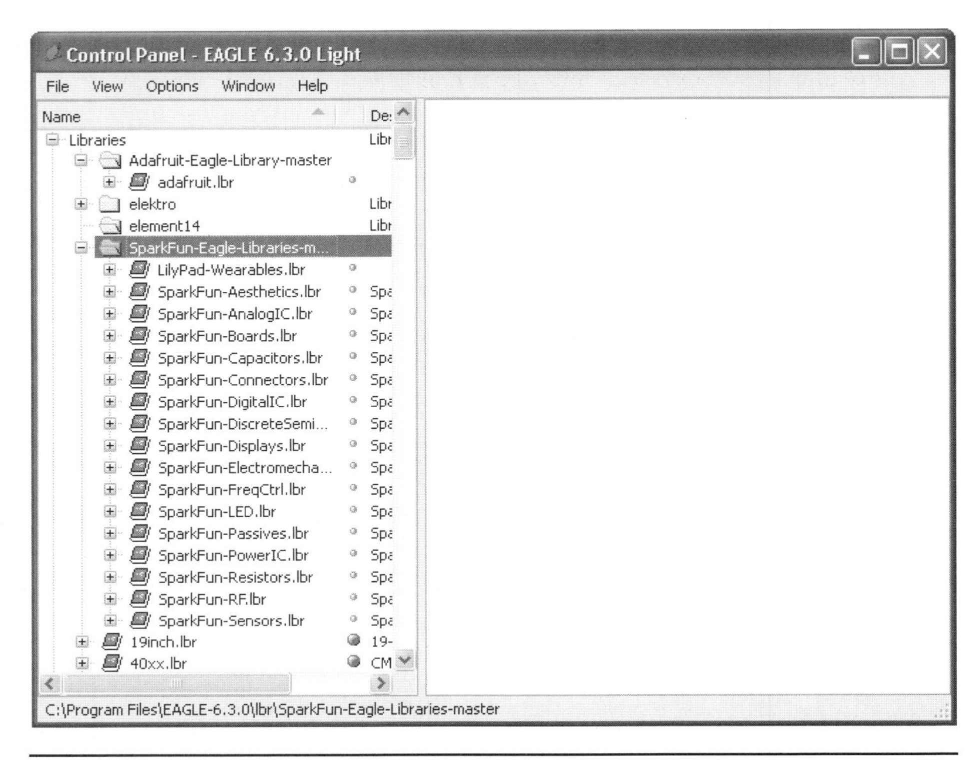

Figure 1-20 Sparkfun and Adafruit libraries installed.

The Sparkfun design rules can be downloaded from the Sparkfun site at www.sparkfun.com/tutorials/115. They are contained in a single `.drc` file. You can right-click on this file on the web page and chose "Save As" to directly save it into the `drc` folder within your EAGLE installation.

Downloading the Book Examples

This book develops a number of example EAGLE projects. You will find it useful to download these so that you can open them up and take a look at them from within the EAGLE software. These files can be downloaded from www.simonmonk.org. Just follow the link to this book.

Light Edition Limitations

The developers of EAGLE have to make some money somewhere, and professionals will soon find that they need features of EAGLE that are not available in the light edition. However, for most of us, the Light edition will do just fine.

There are no restrictions on the number of projects you can design using the software. The main restrictions of the light edition include the following:

- The maximum size of a board that you can design with this edition is 4 × 3.2 in. (100 × 80 mm). This may not sound very big, but actually you can fit a lot of components in an area of that size. Also, because the PCB fabrication services tend to get expensive as boards get bigger, it is no bad thing to keep your designs compact.
- A maximum of two layers is allowed. This means that you can have copper tracks on the top of the board and on the bottom of the board, but you cannot produce four-layer (or more) boards. Again, this is not a problem for hobby use.
- You can only create one sheet on the Schematic Editor. When designing a circuit, electronics engineers will often split a complex design into a number of schematic diagrams with links from one sheet to another. This restriction means that you have to fit all your design onto a single schematic diagram. This is nowhere near as restrictive as it might sound because there is no practical limit to how big that sheet can be.

EAGLE Light can be used as freeware as long as the use is noncommercial or for nonprofit applications. If you plan to build a business out of a product that you design with EAGLE Light, then you should buy a commercial license. For EAGLE Light, this is only $70 at the time of this writing.

Summary

In this chapter you have learned a little about PCBs, installed EAGLE, and run it for the first time. You can now move on to Chapter 2, where you can start designing a simple project using EAGLE.

CHAPTER 2
Quickstart

In this chapter we will get to grips with EAGLE and create the schematic and board files for a simple project that uses a 555 timer to flash a light-emitting diode (LED).

As the chapter title suggests, this chapter is all about doing it quickly rather than doing it with best practice. In later chapters, especially Chapters 4 and 5, you will find a more thorough and considered approach to design. Seasoned electronic engineers might need to avert their eyes for some of this chapter.

Creating a New Project

The first step is to create a new project. Thus, from the "File" menu, select "New" and then "Project" (Figure 2-1). This will allow you to edit the name of the project (Figure 2-2).

You now need to create a new schematic diagram so that you can start drawing out the circuit. To do this, right-click on the "Flasher" project icon and select "New" and then "Schematic" from the popup menu (Figure 2-3).

Initially, the schematic document will be labeled "Untitled." It is a good idea to save it by selecting "File" and then "Save As" from the menu. Give it the name `flasher.sch`. You can actually have more than one schematic and board file within the same project. However, this can lead to confusion, and it is easier to just have one of each and give the files the same name as the project. I like to use an uppercase initial letter for the project and lowercase for the files.

Do not create a board file yet. We will do that after you have designed the schematic.

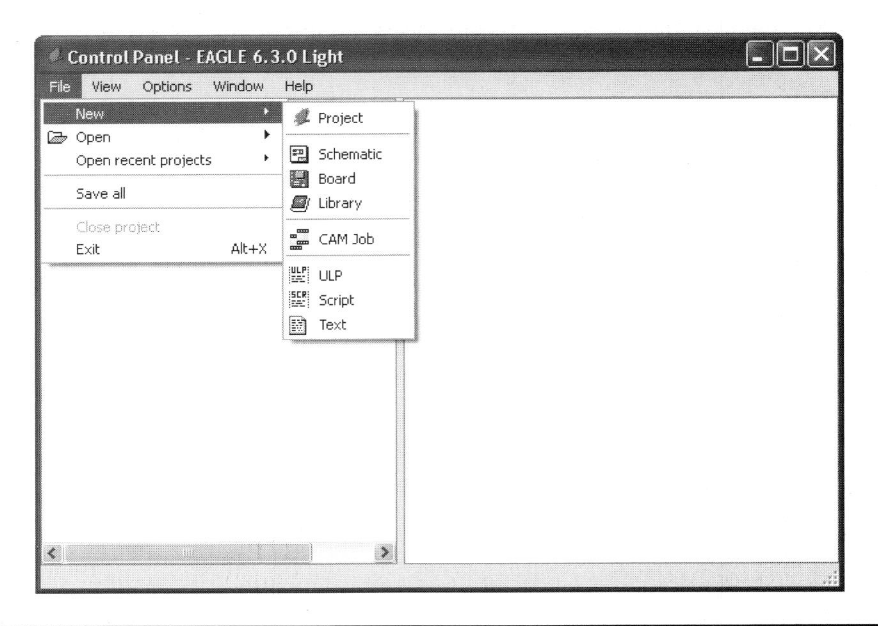

FIGURE 2-1 Creating a new project.

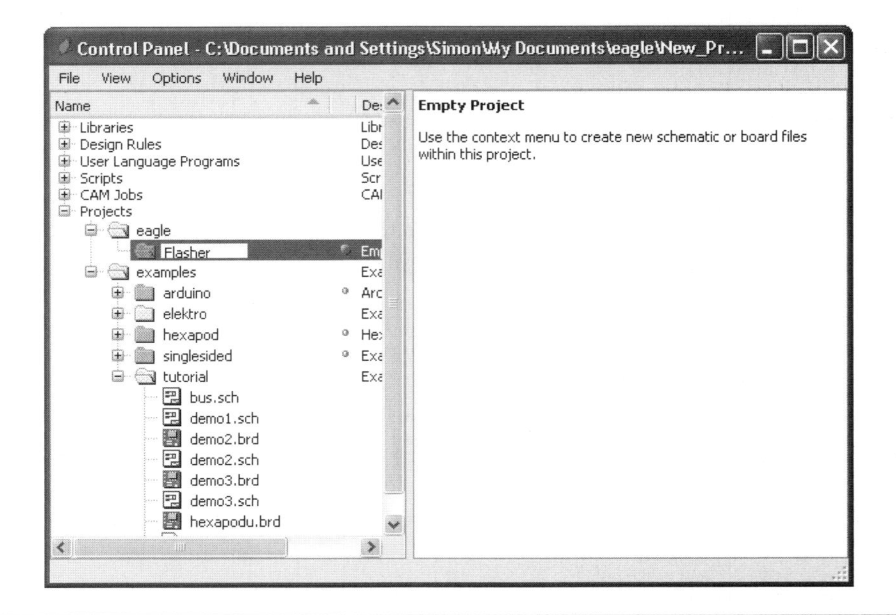

FIGURE 2-2 Naming the new project.

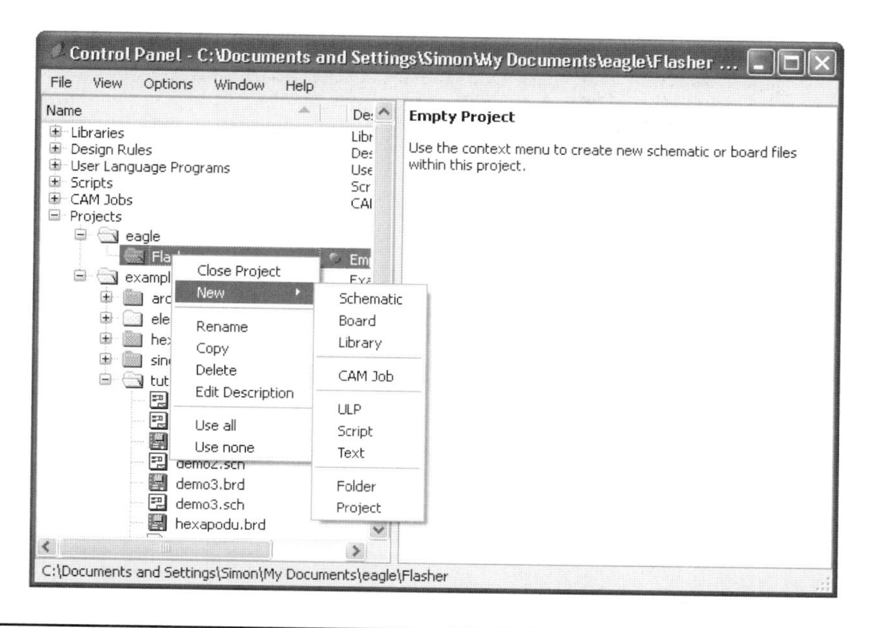

FIGURE 2-3 Creating a new schematic.

Figure 2-4 shows the schematic window. As you can see, there are a bewildering number of sometimes cryptic icons on the left. If you hover the mouse over them, you will get some indication of what they do in the status area at the bottom of the window.

In Figure 2-4, the most important icons have been highlighted. The "Add" icon is used when adding new components to the schematic, and as you might expect, the "Delete" button removes them. Less obviously, there is also an icon for "Move" as well as icons for "Name" and "Value" that will change the properties of the components that have been added.

EAGLE operates on a different principle than most document editing systems. As an example, in a word processor, you generally would highlight some text and then click on the "Bold" icon to make the text bold. You select the thing or things and then select an action to apply to them. This is not the case for EAGLE. In EAGLE, it is the other way around. First, you select the action (Move, Delete, Name, Value, Net, etc.), and then you select the component you want to apply it to. This is a little confusing at first, but after a while, it will become second nature.

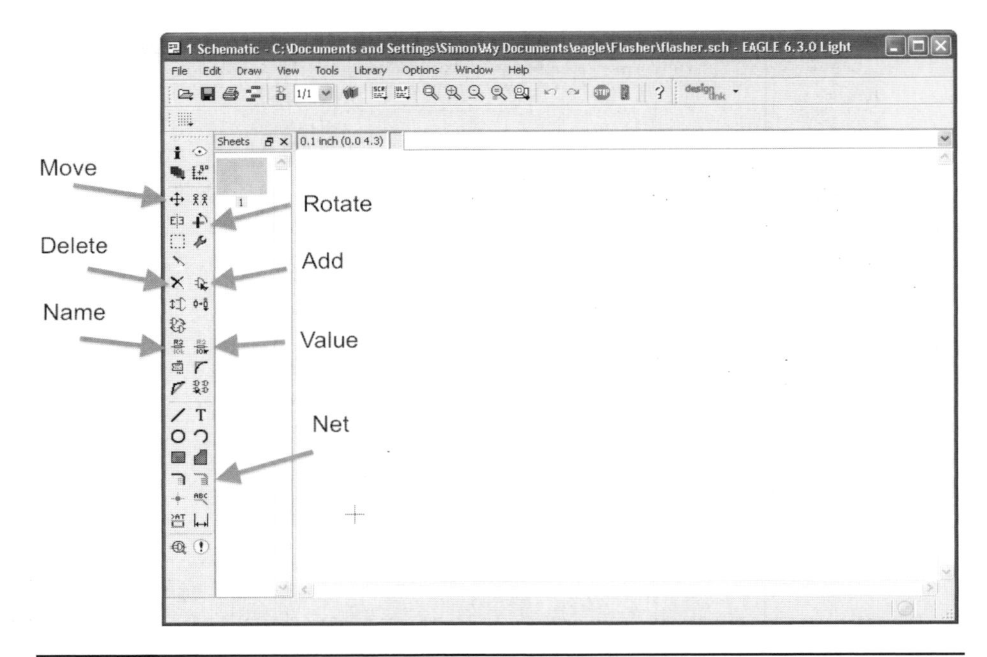

Figure 2-4 Schematic window.

Drawing the Schematic

You are now ready to start drawing the schematic diagram for this project. This is really a two-stage process. For a simple project such as this, it is easiest to add all the components and then start connecting them together.

Add the Components

This project is a simple LED flasher that uses the 555 timer integrated circuit (IC). Because this is the focus of the project, start by adding a 555 timer by clicking on the "Add" button. Having clicked on "Add," you will be greeted with the vast array of libraries shown in Figure 2-5.

We need to find a 555 timer IC somewhere in there. You can do this by searching. Start by entering "555" in the search field. The result is shown in Figure 2-6.

The result is only one component, and it's not the right one. It's a 556, which is the dual-timer version of the 555. The problem is that the search uses entire words. To find any words that end in "555," add an asterisk (*) to the front of the search. Now we have found a good selection of devices. Clicking on a device

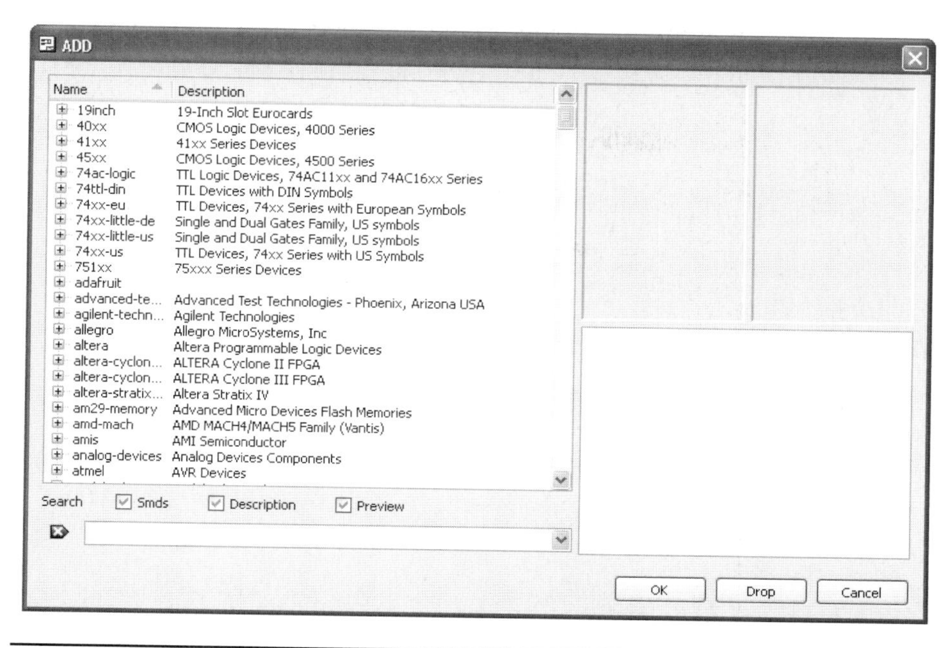

Figure 2-5 Finding a component.

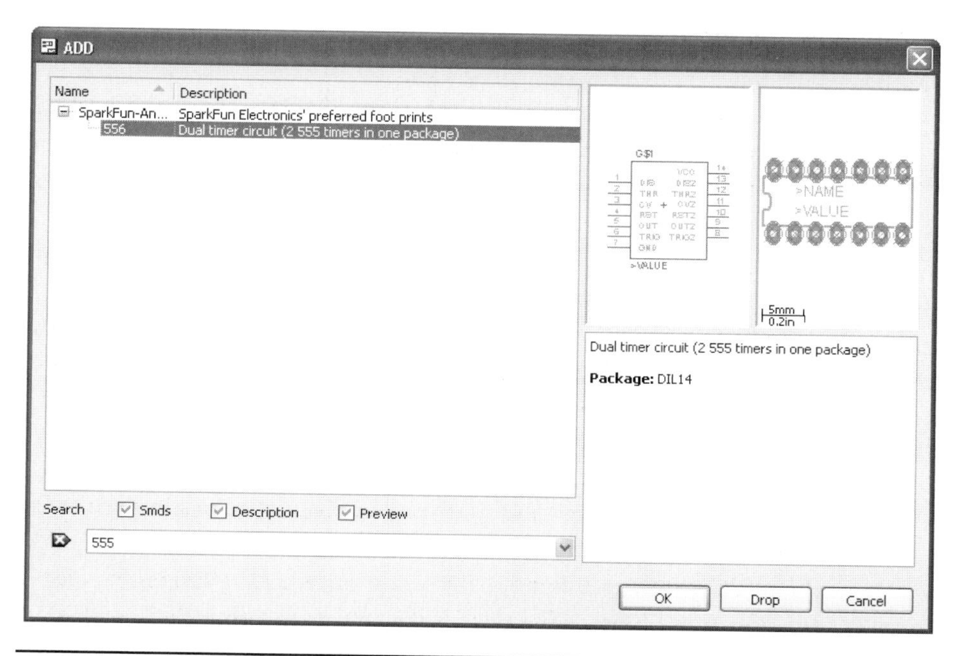

Figure 2-6 Search results.

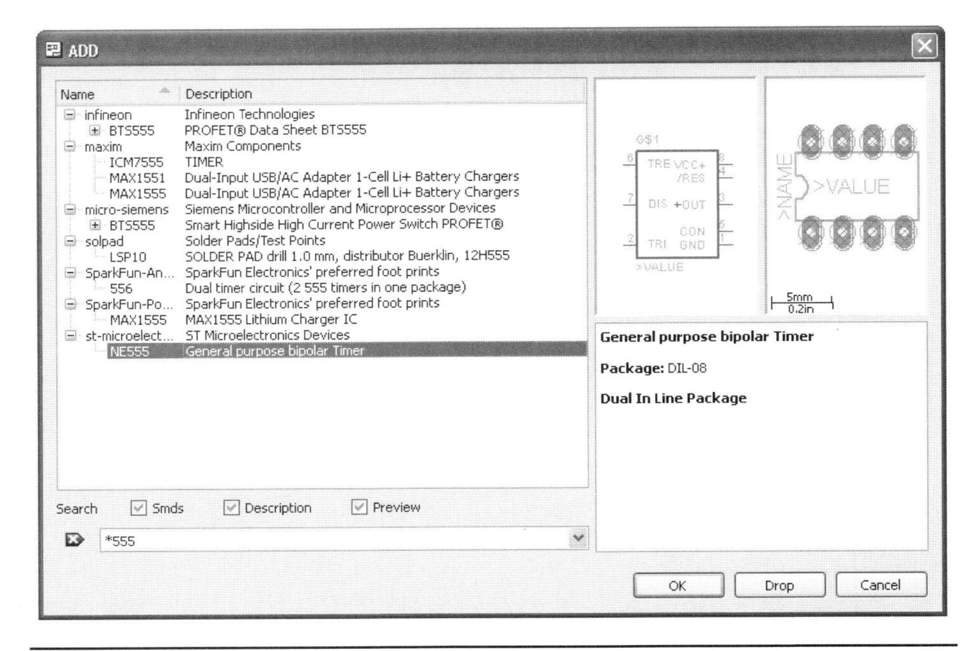

FIGURE 2-7 Wild-card search.

shows a preview of it. Thus you can see the NE555 we are looking for highlighted in Figure 2-7.

Click "OK," and move the cursor toward the middle of the schematic window. Then click to drop the component on the schematic. Note that if you click several times, you will end up with several copies of the component. Make a couple of extra clicks so that you can practice deleting things (Figure 2-8).

Note that to escape from component adding mode, you need to press the "ESC" key and then click on "Cancel" in the "Add" window.

Remember that to delete the extra components, you first need to switch to "Delete" mode by clicking on the "Delete" icon. Until you change modes again, everything you click on will be deleted. So click on all but one of the NE555 components (Figure 2-9).

Time to add some new components. Let's add the resistors. When adding components such as resistors, you do not find resistors of any particular value listed in the component libraries but rather components of a particular type and size. We are going to go looking for ¼-W through-hole resistors.

Searching the libraries for resistors is going to bring back rather a lot of answers. Thus, to reduce that number, uncheck the box that says "SMDs" (Figure 2-10).

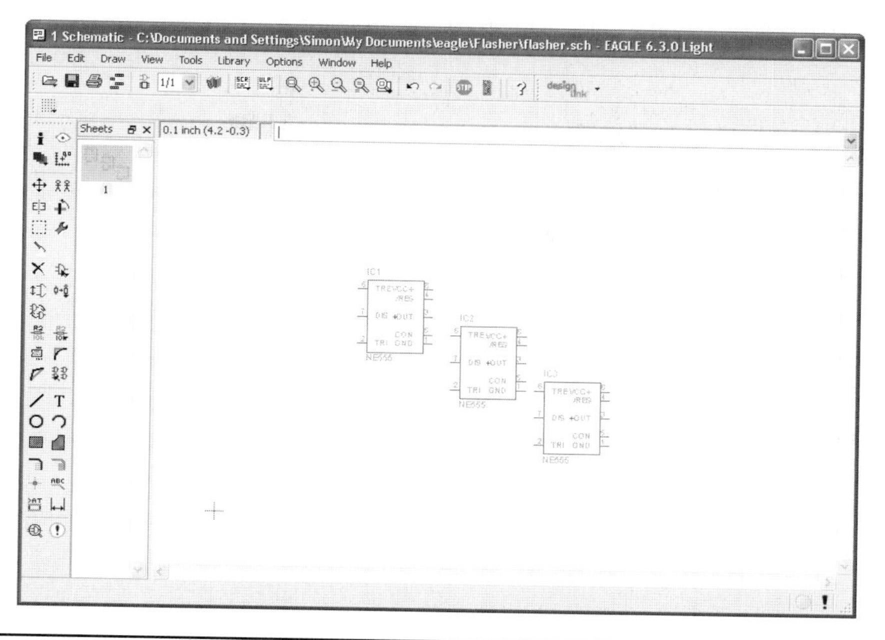

FIGURE 2-8 Too many 555s.

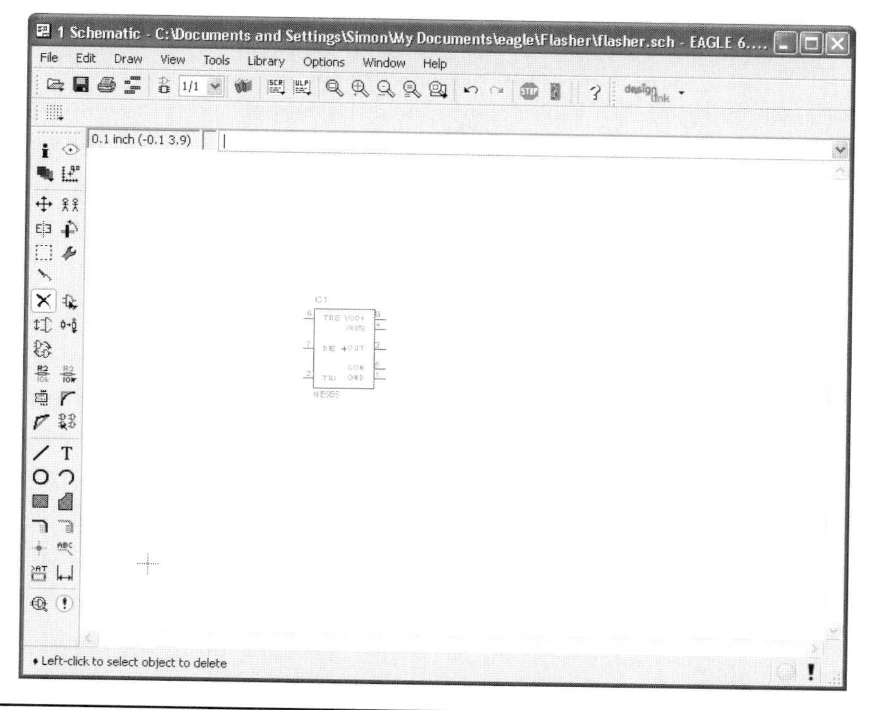

FIGURE 2-9 Deleting components.

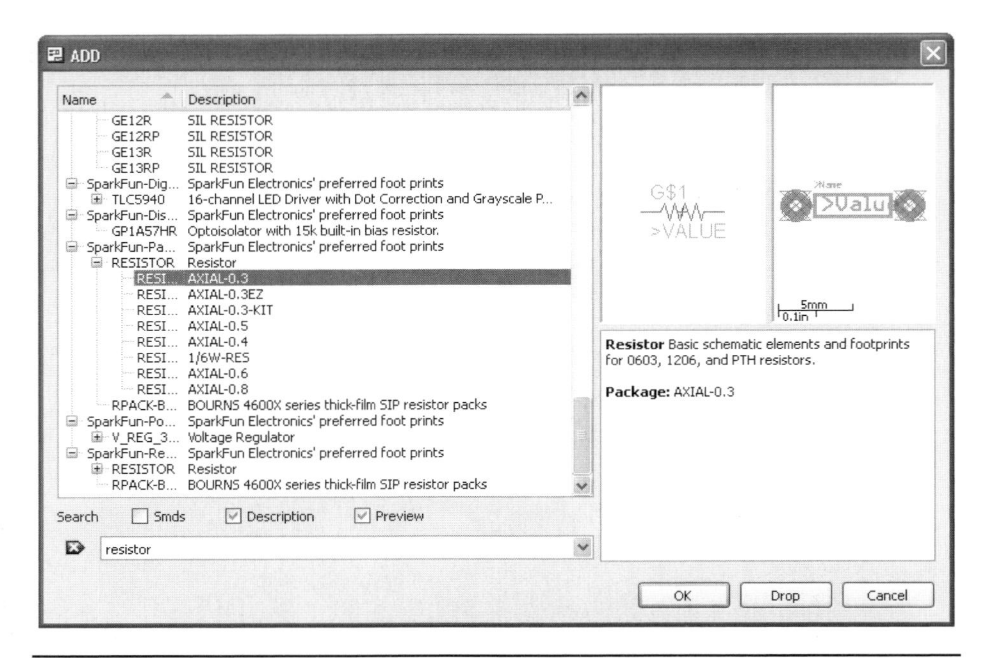

FIGURE 2-10 Finding resistors.

There are still a lot of results, and it probably does not matter exactly which one we select. Many of them will be nearly identical. Scroll down to the Sparkfun results, and pick the resistor option shown in Figure 2-10. You will need to drop four of these onto the schematic window (Figure 2-11).

It does not matter yet where you drop them; later on, you will have to move everything around anyway.

Now add an electrolytic capacitor to the schematic. It's going to be a 1-μF capacitor, so it can be one of the smaller devices. Search for "electrolytic capacitor," and scroll down to the Sparkfun section again. The "CAP_POL" sections are for polarized or electrolytic capacitors, so look in there. There isn't a 1-μF capacitor, but there is a 10-μF one (CPOL-RADIAL-10UF-25V) that will do just fine.

Add two LEDs as well (search for "LED5mm"), and select the one from the Adafruit library.

When all of these components have been added, your schematic will look something like Figure 2-12.

The final part that we need to add to the schematic is not really an electronic component but some means of supplying power to the circuit. Let's use a screw terminal. Even if we just end up soldering a battery clip to the pads, at least we have two convenient pads to solder to.

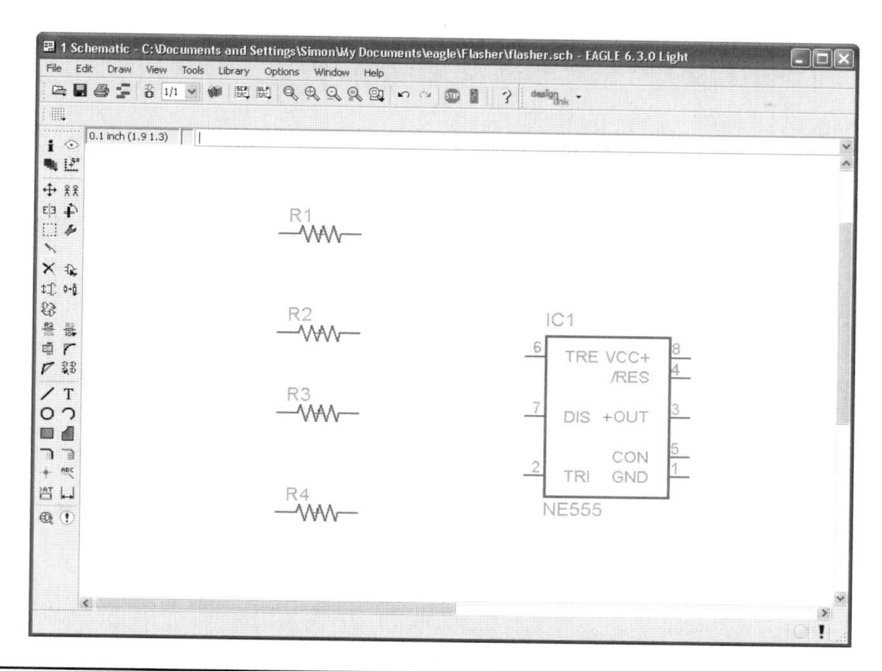

FIGURE 2-11 Adding resistors to the schematic.

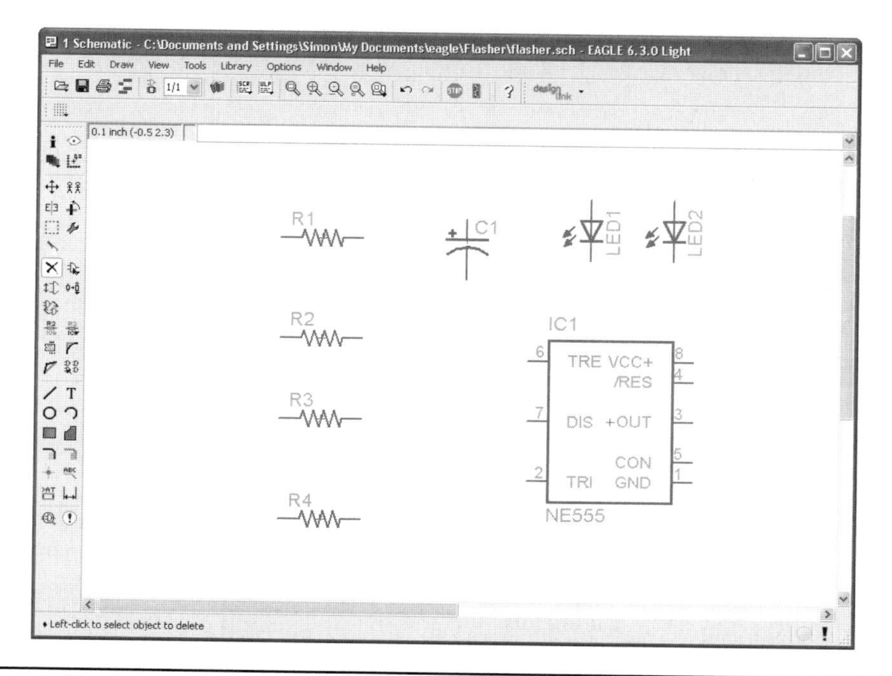

FIGURE 2-12 Schematic with all the components.

Search for and add a two-way screw terminal. Use "terminal" as the search term, and select the top result from Adafruit. Your schematic should look like Figure 2-13.

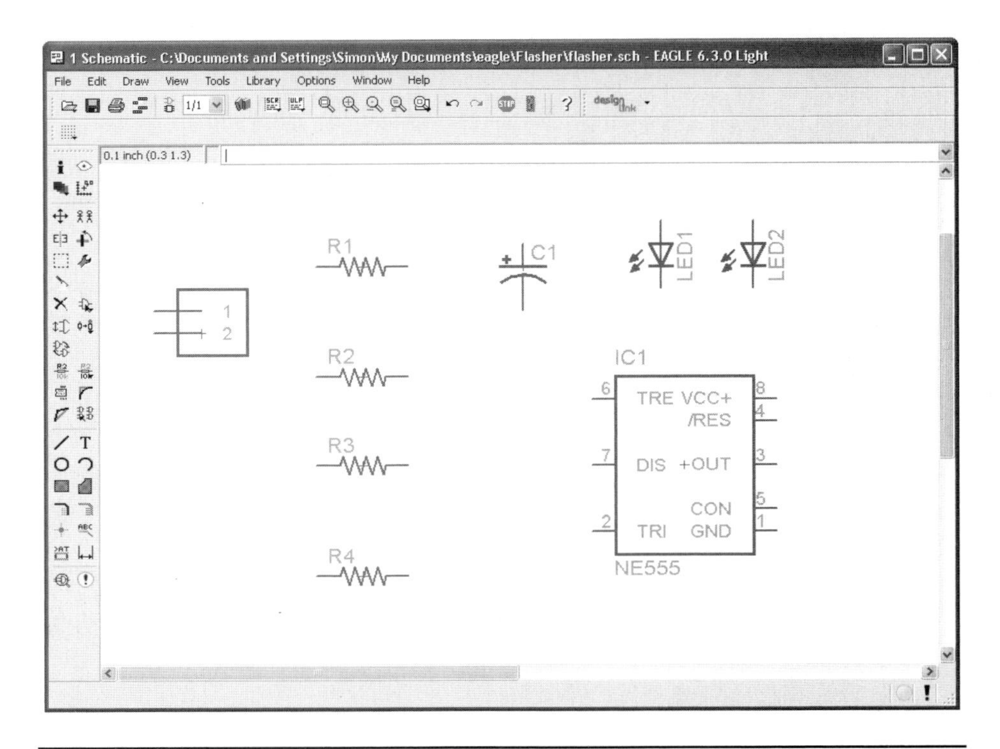

FIGURE 2-13 Schematic with screw terminals added.

Now that all the components have been added, we can rearrange them so as to be ready to connect them up. To do this, we will need to change the orientation of some of the components. In particular, all the resistors need to be vertical rather than horizontal, and the power connector needs to be rotated through 180 degrees so that the connections face toward the rest of the components.

Rotating components, like everything else in EAGLE, requires you to switch to a particular mode. Select the "Rotate" mode by clicking on the "Rotate" icon (see Figure 2-4). Click on each of the resistors to rotate them through 90 degrees, and then click on the connector twice to rotate it a full 180 degrees (Figure 2-14).

To move the components, select the "Move" button, and drag the components to their approximate locations, as shown in Figure 2-15.

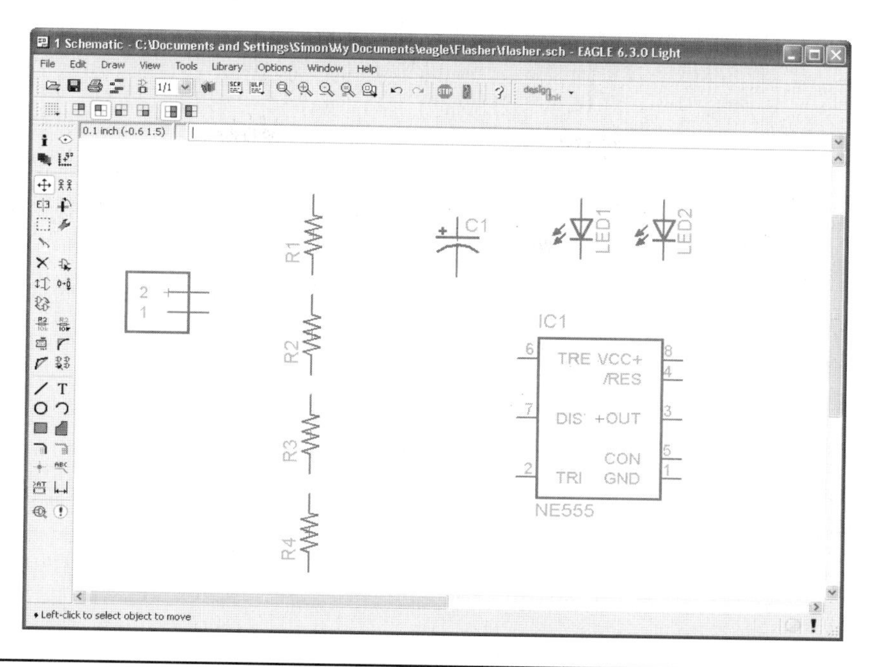

FIGURE 2-14 Rotating components.

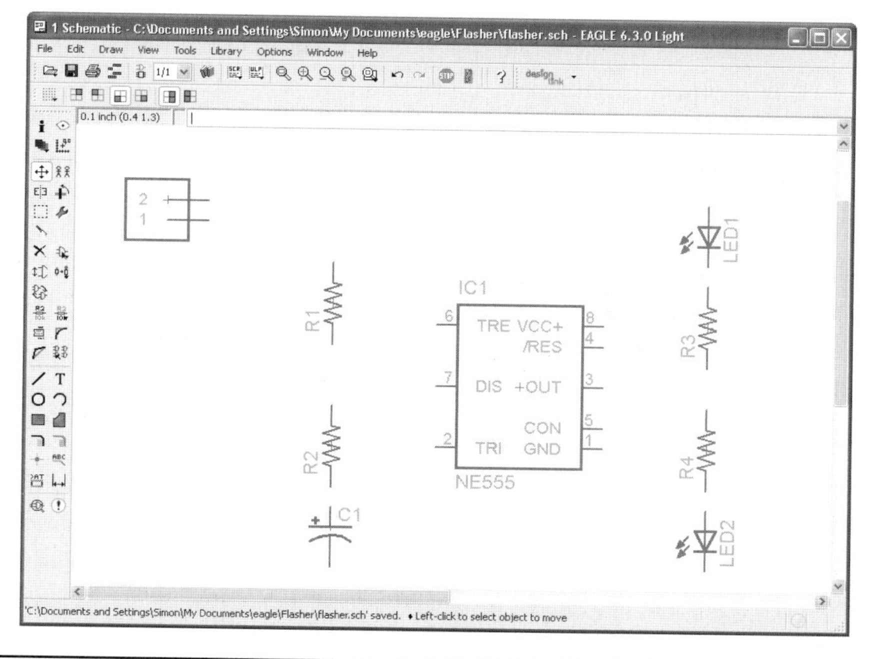

FIGURE 2-15 Moving components.

Join Them Together

At last, the time has come to start connecting the components together. To do this, we need to click on the "Net" icon (see Figure 2-4). Do not be fooled into using the "Wire" icon; this is something entirely different.

To attach one component connection to another, click on the lead of the component you want to attach to, and a green line will appear. You can then click again anywhere you want the line to bend through 90 degrees until you come to some other component lead, where the connection will finish. As an easy first one, connect the bottom of R3 to the top of R4, and then add in the rest of the connections until your schematic looks like Figure 2-16.

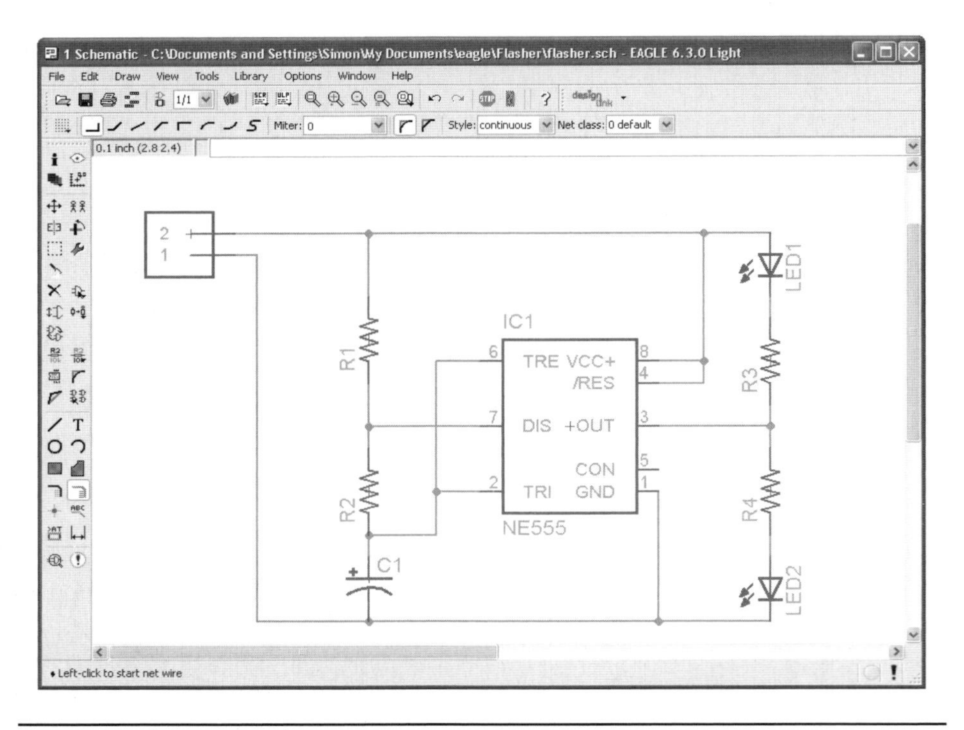

Figure 2-16 Connecting the components.

So far, so good, but none of our components have any values marked for them. To add values, click on the "Value" icon to put EAGLE into value-setting mode. Go through each of the components, setting their values as shown in Figure 2-17.

This is now our completed schematic. In later chapters we will find out more things that we can do with the schematic, but for now, this nice, simple diagram is good enough.

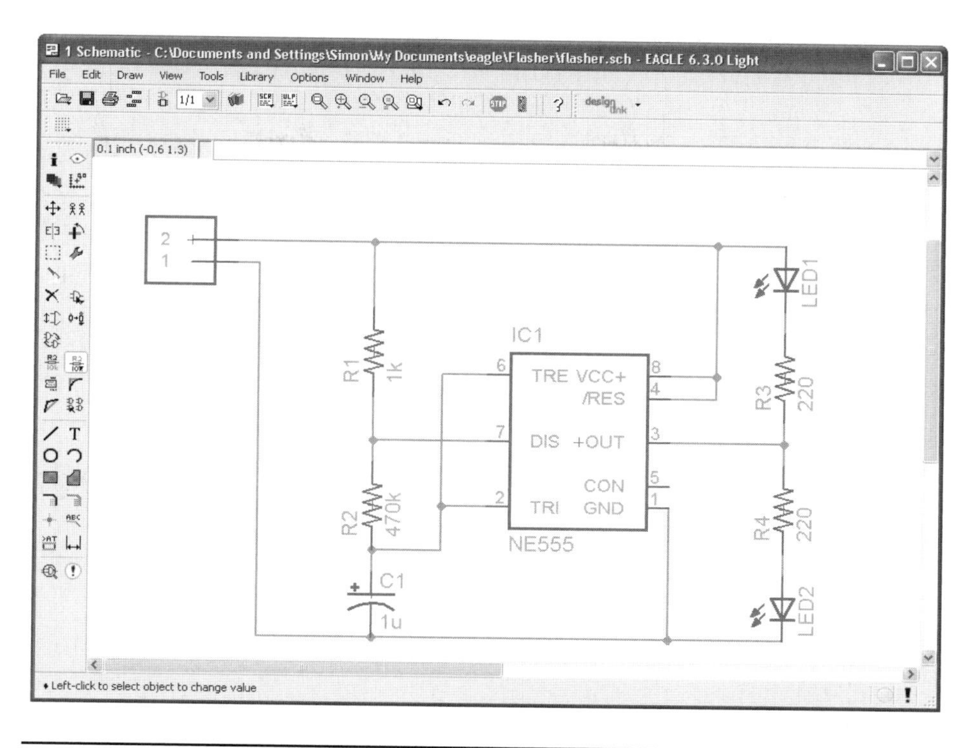

Figure 2-17 Setting component values.

Electrical Rule Check

Before we start laying out the board, we should run an *electrical rule check* (ERC). This will analyze our design and tell us if there are any problems with it. To do this, from the "Tools" menu of the "Schematic" window, select the option "ERC." The result will be something like Figure 2-18.

The first thing to note is that only one of the problems listed is an error; the other four are warnings. The error is because we have left pin 5 of IC1 unconnected. The part specification for pin 5 says that this is not allowed. We could just ignore the error, but checking the datasheet for the 555 timer, it is normal to attach a 10-nF capacitor between pin 5 and ground (GND), so do that now. Search for "capacitor," and then find CAP-PTH-SMALL in the Sparkfun CAP section.

Add it, connect it up, and change the value to be 10 nF, and your schematic now should look like Figure 2-19.

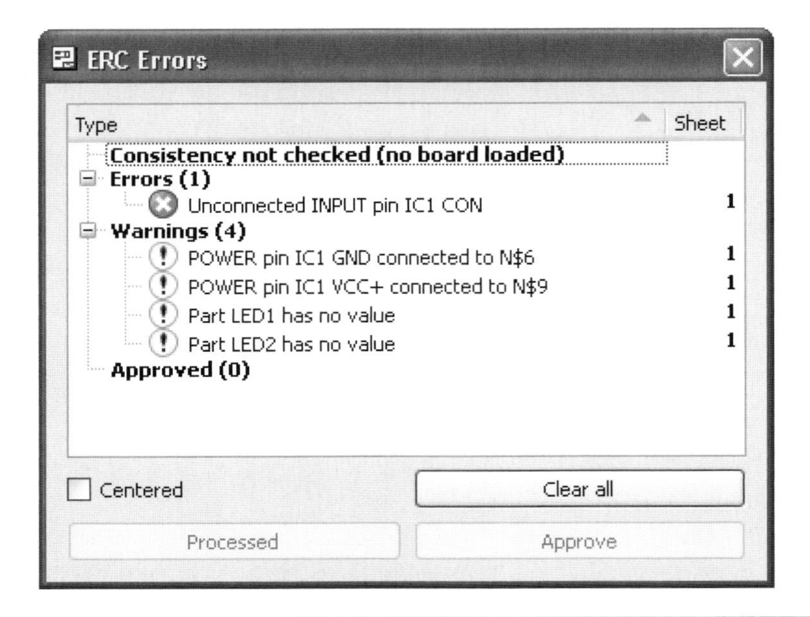

FIGURE 2-18 Results of an electrical rule check.

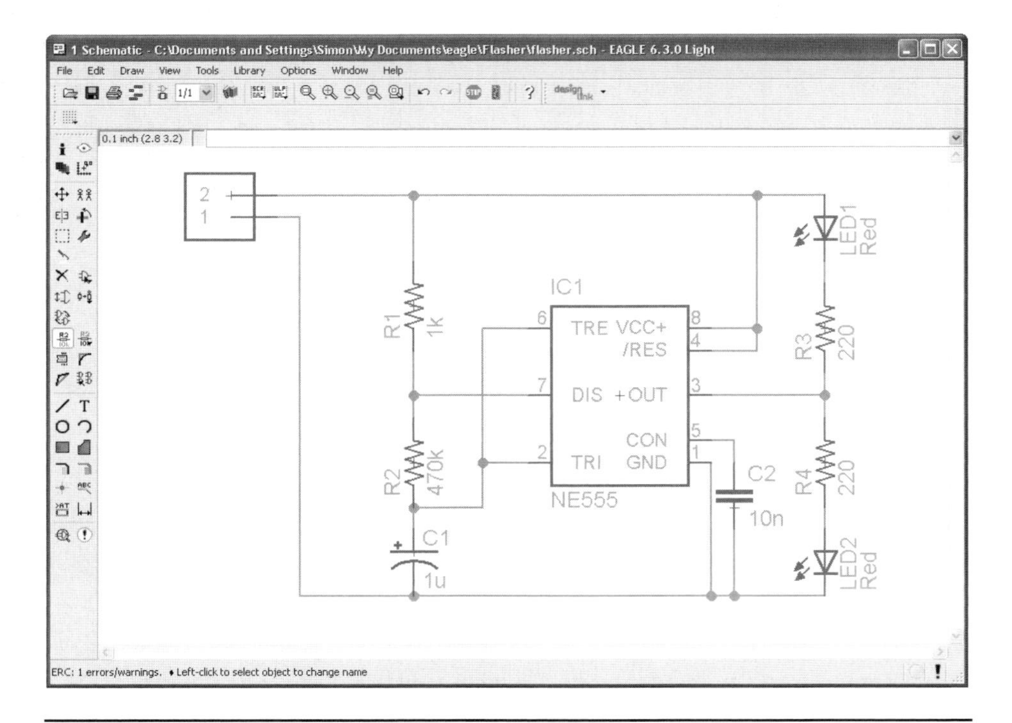

FIGURE 2-19 Schematic with capacitor for pin 5 added.

Close the "ERC" window, and run it again. Just the four warnings should remain. The last two of these just say that the LEDs do not have a value. We can either ignore these warnings or add a value of say `Red` to indicate the color of the LEDs.

This just leaves us with two warnings. They will say something like `POWER pin IC1 GND connected to N$6`. The N$6 probably will be different because it depends on the order in which you connected things together. N$6 is an automatically assigned name for a net. A *net* is, if you like, a line, and all the lines that connect to it between one component pin and another.

The warning is there because IC1 expects its GND pin to be connected to a net called `GND`, but my ground net is called `N$6`. You can discover the names of your nets by clicking on the "Show" icon in the "Tools" menu (at the top; it looks like an eye). Clicking on a net will then highlight it and show you its name in the status area at the bottom of the screen (Figure 2-20).

This is easy enough to remedy. GND is a much better name for the net than `N$6` anyway, so select the "Name" tool and then click on the "N$6" net and change its name to `GND`.

Run the ERC again, and you should just be left with a warning like this: `POWER pin IC1 VCC+ connected to N$9`. This is very similar to the complaint about the GND connection but is for the positive power supply to the IC (VCC+). Once again, anything for a quiet life, so change the name of the net to `VCC+` in the same way as you did the GND net.

Now, when you run the ERC, no window will appear; you will just see a message in the status area at the bottom of the screen that says `ERC: No errors/warnings`.

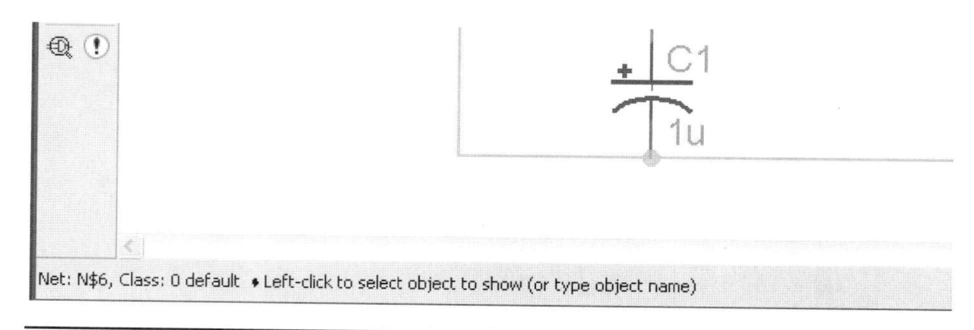

FIGURE 2-20 Using the "Show" tool.

Laying Out the Board

You can now start creating a board layout. The most convenient way to do this is to select the option "Switch to Board" on the "File" menu of the Schematic Editor. This will result in a prompt saying (and I paraphrase), "There is no board, so would you like to create one from the schematic?" This is just what we want, so click on "Yes."

Initially, our board does not look very promising (Figure 2-21). Clearly, we still have some work to do.

Dragging Components onto the Board

All the components are bunched up in the bottom left of the screen, and there is the outline of a wire rectangle to the right. This rectangle represents the borders of the circuit board itself. EAGLE has not presumed any initial layout, so the first thing we must do is to use the "Move" tool to drag all the components onto the board.

You will notice that all the legs of the components have yellow lines attaching them to each other. These are called *air wires*. They indicate a connection that at

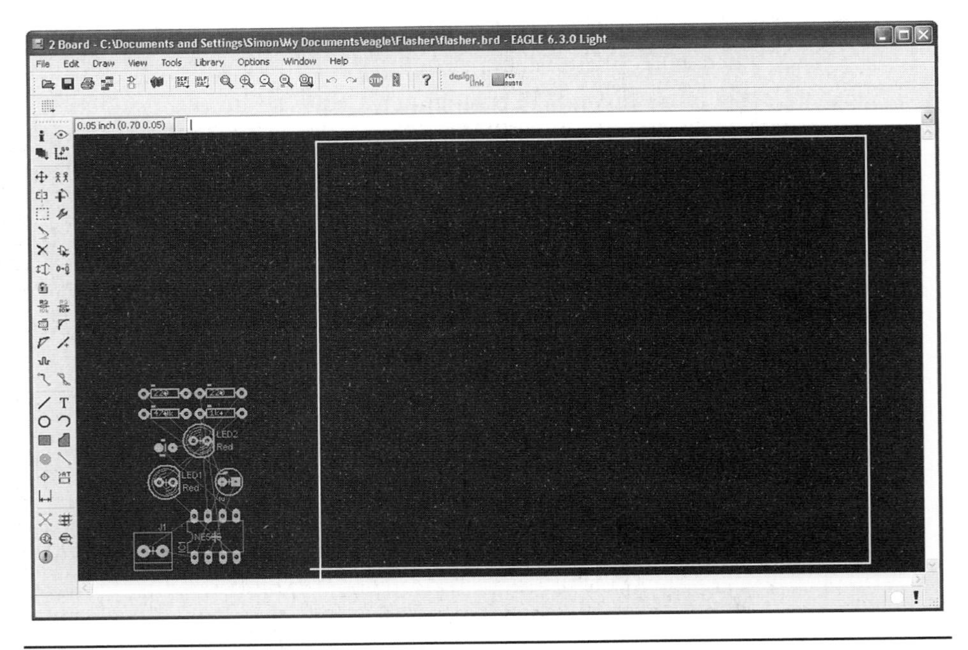

FIGURE 2-21 Board layout editor.

some point we will have to convert into tracks that replace the air wire with a real connection. They do, however, help us to decide where to place the components so as to minimize the crossing of these lines.

This board is actually much bigger than we need for this simple project, but it is often easier to at least place all the components before adjusting the size of the board. The board has an origin at the bottom left; as you move your mouse cursor around the board area, you will see some numbers change just next to the "Show" (eye) icon. These coordinates are in inches. Because through-hole components generally conform to a 1/10-in. pin spacing, this is a more convenient unit to use than metric units.

Select the "Move" tool, and move the components onto the PCB area. Then use the "Rotate" tool so that the components look something like Figure 2-22. Note that you will not be allowed to rotate a component until it is on the board. This is just one of EAGLE's little quirks.

Positioning and rotating the components are very much a matter of trial and error. Every so often, click on the "Ratsnest" button near the bottom left of the "Tool" pallet. This will redraw the air wires, keeping them as short as possible. You can zoom in and out either using the "Zoom" buttons on the toolbar or using the scroll wheel on your mouse.

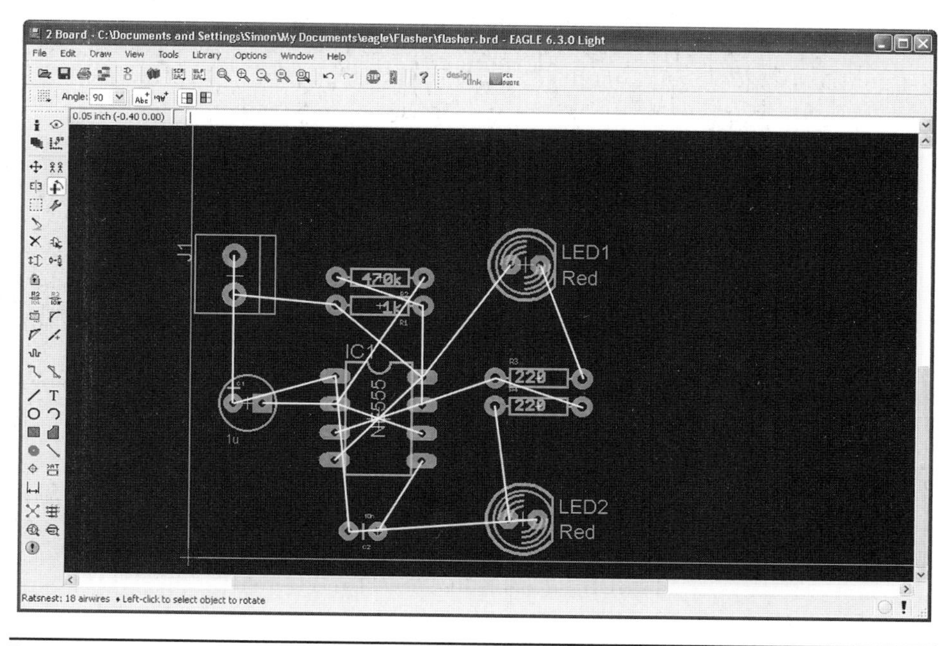

FIGURE 2-22 Positioning the components.

Resizing the Board

Our components are only occupying the bottom-left quarter of the board, so now is a good time to make the board smaller. Zoom out so that you can see the whole board, and then select the "Move" tool and drag in the left and top sides of the board until they just enclose all the components (Figure 2-23).

Routing

The next stage of the process is to route the tracks. EAGLE can automatically route your design for you, and for a simple design like this, we might just as well let it. In Chapter 5, we will look at laying out PCBs by hand.

To start the "Autorouting" tool, click on the "Auto" button in the "Tool" palette. The icon looks a bit like a grid and is immediately to the right of the "Ratsnest" icon. This will launch the autorouting dialog (Figure 2-24).

We can come back to what all these options mean in Chapter 5. For now, though, just accept the defaults and click the "OK" button. The board will be routed before your very eyes. If the routing is successful, then the status bar will say `Ratsnest: Nothing to do`. If for some reason the autorouter was unable to eliminate all the air wires, then the message would tell you how many air wires remained. You could then route them manually or change the parameters in the autorouting dialog and try again. Figure 2-25 shows the final layout of the board.

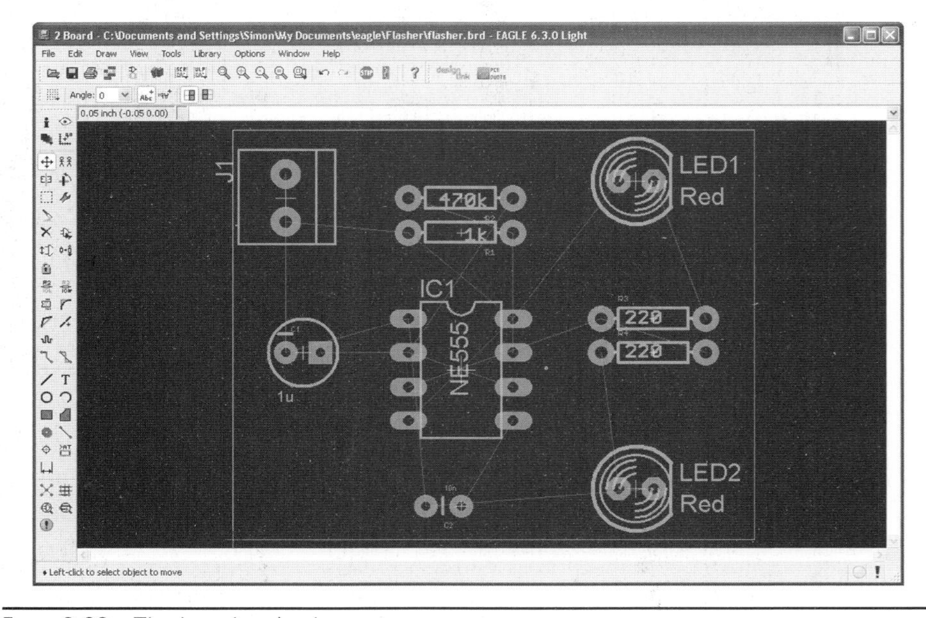

FIGURE 2-23 The board resized.

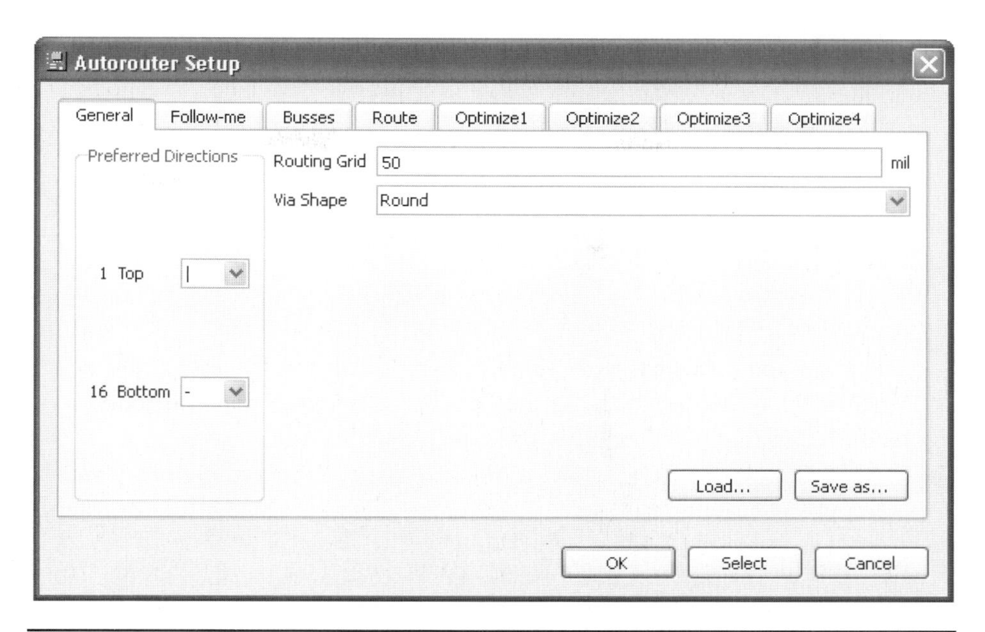

FIGURE 2-24 Autorouting dialog.

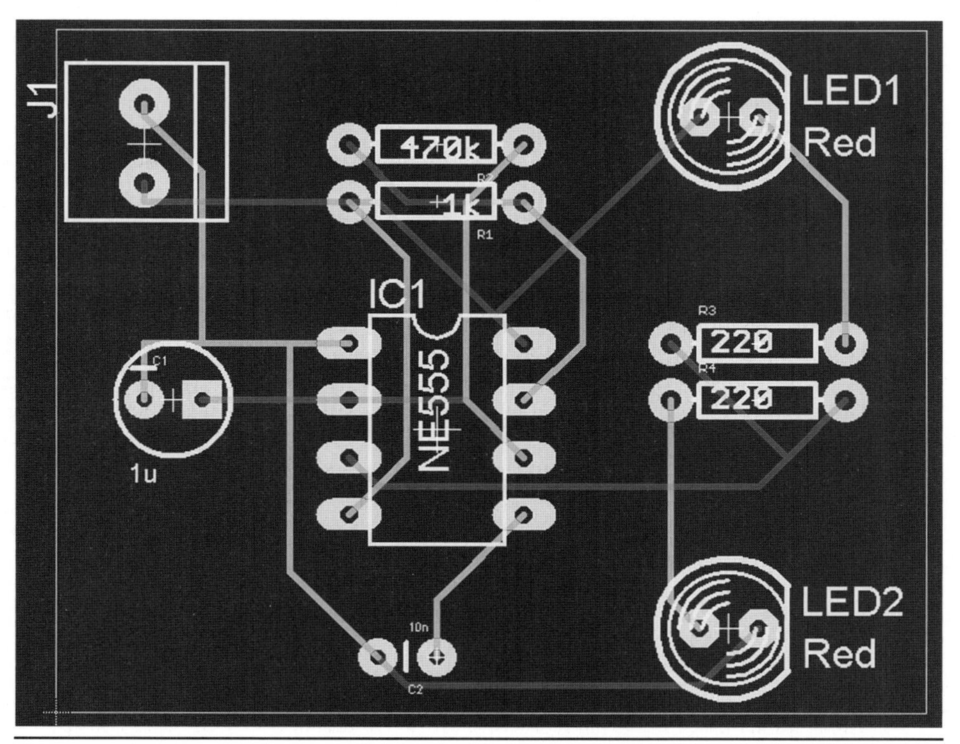

FIGURE 2-25 Final board layout.

If you have been following this in EAGLE as you read or are reading a color edition of this book on an e-book reader, then you will notice that the board is starting to look quite colorful. The red lines represent tracks on the top of the board, and the blue lines track on the bottom. The pads are shown in bright green, and the white lines show the silkscreen layer.

Summary

This chapter has shown you just enough to draw a schematic and then lay out a board from the design. By design, a lot of material and background have been omitted so that you can at least get started with EAGLE.

In Chapter 3, we will take a more detailed look at components and libraries because identifying the right parts for your design can be time-consuming and difficult.

CHAPTER 3

Components and Libraries

Finding the right components for a project can be a time-consuming process. If you are using EAGLE, then a further complication is that you need to either use components that are already in the EAGLE libraries or download a library that includes the part or, as a final resort, create your own part and add it to a library. This chapter serves as a reference for the most common components that are used by hobbyists, as well as showing you where to find EAGLE models for components and even the components themselves.

U.S. versus European Circuit Symbols

When you are choosing a component from a library to use in a schematic, the most important thing about it relates more to the board layout than to the schematic. If it does not have pads in the right places, then it will not be of any use. Another consideration is how the symbol is drawn on the schematic. Unfortunately, there is more than one standard for component symbols in use. The main divide is between symbols commonly used in the United States and those in use in Europe.

If you browse through the libraries, you will often find two versions of each component. For example, R-EU and R-US for European and U.S. resistor symbols. Figure 3-1 shows the symbols for resistors and capacitors.

In this book, we will stick to the U.S. circuit symbols largely because the useful Adafruit and Sparkfun libraries are in this format. If you feel strongly about using the European-style symbols, then you can search out circuit symbols in that standard.

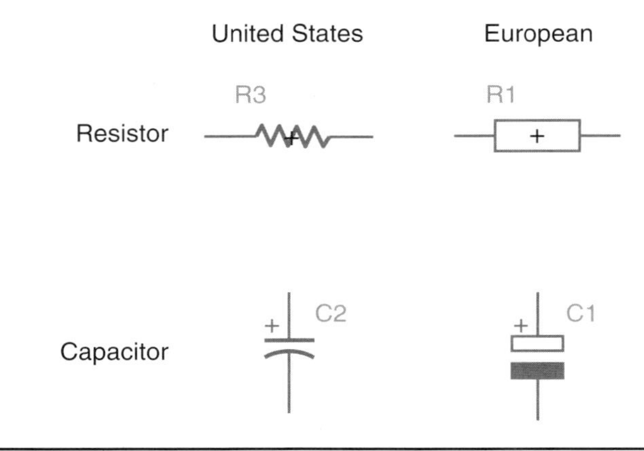

Figure 3-1 United States and European circuit symbols.

Resistors

Resistors are probably the easiest components to find in a library. They are pretty standard in size, with relatively few different sizes to choose from, both in through-hole and surface-mount device (SMD) forms.

Through-Hole Resistors

For most though-hole designs, ¼-W metal-film resistors are fine. These normally will be mounted flat against the PCB with the leads bent at right angles at the ends of the resistor bodies. This usually requires a 0.3-in. separation. Figure 3-2 shows a selection of through-hole resistors. The resistor power ratings from front to back are 125 mW, ¼ W, and two 1-W resistors.

Table 3-1 details common resistor sizes (from wattage ratings). These are taken from the Sparkfun library. You can find them in the library by searching using the search term "RESISTORPTH-*."

The easiest way to identify any component is to keep it in a component box that labels exactly what it is. If, however, the components get mixed up, then through-hole resistors are identified by their color-coded stripes.

Each color has a value as follows:

Black	0
Brown	1
Red	2
Orange	3

Yellow	4
Green	5
Blue	6
Violet	7
Gray	8
White	9
Gold	1/10
Silver	1/100

Besides representing the fractions 1/10 and 1/100, gold and silver are also used to indicate how accurate the resistor is, so gold is ±5 percent and silver is ±10 percent.

There generally will be three of these bands together starting at one end of the resistor, a gap, and then a single band at the other end of the resistor. The single band indicates the accuracy of the resistor value.

Figure 3-3 shows the arrangement of the colored bands. The resistor shown uses just the three bands. The first band is the first digit, the second band is the second digit, and the third "multiplier" band is how many zeros to put after the first two digits.

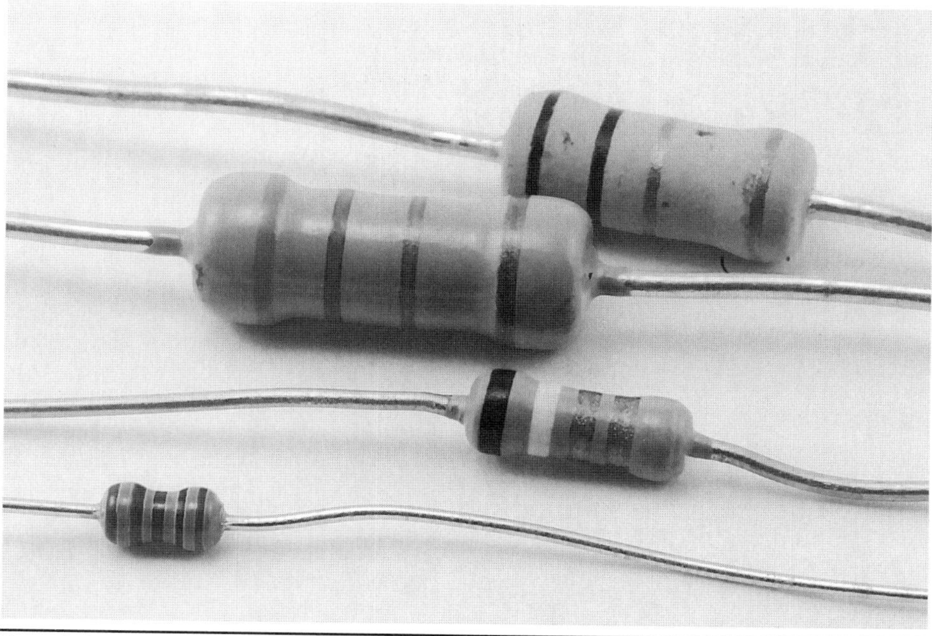

FIGURE 3-2 Through-hole resistors.

Table 3-1 Through-Hole Resistors: Common Component Sizes

Wattage	Library	Footprint
⅛ W (125 mW)	RESISTORPTH-1/8W	>NAME >VAL
¼ W (250 mW)	RESISTORPTH-1/4W	>Name >Value
½ W	RESISTORPTH-1/2W	>Name >Value
1 W	RESISTORPTH-1W	>Name >Value
2 W	RESISTORPTH-2W	>Name >Value

A 270-Ω resistor will have a first digit of 2 (red), a second digit of 7 (violet), and a multiplier digit of 1 (brown). Similarly, a 10-kΩ resistor will have bands of brown, black, and orange (1, 0, and 000).

Some resistors have four bands rather than three, in which case the first three stripes represent the value, and the last stripe is the multiplier. Thus a 10-kΩ resistor with four stripes would have stripes of brown, black, black, and red.

SMD Resistors

In general, through-hole devices are often of much higher power than necessary. Often ¼-W devices are used as a standard that will do for almost any application.

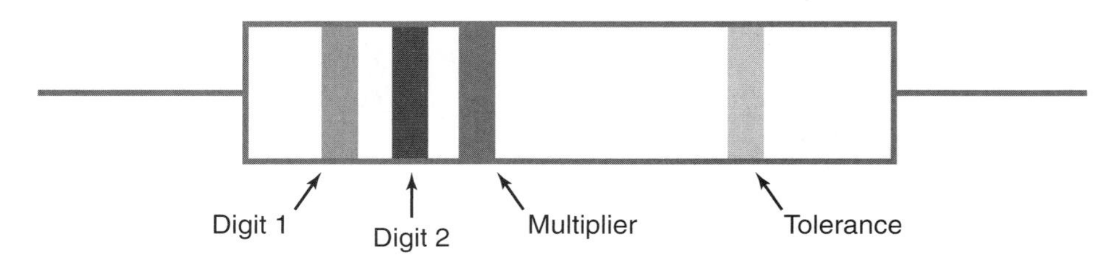

Figure 3-3 Resistor color bands.

SMD resistors are available with a wide range of power ratings, and generally, lower-power SMD resistors are used rather than their through-hole counterparts. These devices start really small. In fact, generally, they are too small to solder by hand.

SMD devices (both resistors and many other two-legged devices) come in standard sizes, denoted by four digits, for example, 0402, 0805, and 1206. Figure 3-4 shows a selection of SMD resistors. From left to right, the two resistor sizes are 0805 and 1206. As you can see from the match head, theses are pretty tiny, and if you are soldering by hand, you should not consider devices smaller than 0805.

The four digits of the package size actually specify the dimensions of the device. The first two digits are the length, and the second two are the width. In both cases, the measurements are in 1/1,000 of an inch. Thus a 0402 device has a length of 4/1,000 of an inch and a width of 2/1,000 of an inch.

Now is a good time to introduce the unit called the *mil*. A mil is not to be confused with a millimeter (mm). A mil is 1/1,000 of an inch and is still used widely in the electronics industry. Things get confusing because in electronics you will find some things in American measurements (mils and inches) and other things in metric units (e.g., 5-mm LEDs). When you come to laying out your board, you will use mils as the main unit, and most through-hole components have lead spacings in mils, often 100 mils (0.1 in.).

Table 3-2 shows the most common SMD resistor sizes.

For higher powers than 1 W, through-hole resistors normally will be used for their better air circulation to allow heat to dissipate.

SMD resistors are generally marked with a four-digit code. This is rather like the color code used in through-hole resistors but just written as digits. Thus both

FIGURE 3-4 SMD resistors.

Table 3-2 Common SMD Resistor Packages

Wattage	SMD Package	Library	Footprint
63 mW	0402	*0402-RES*	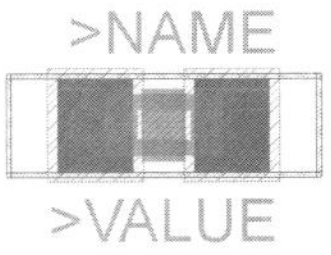 >NAME >VALUE
63 mW	0603	*0603-RES*	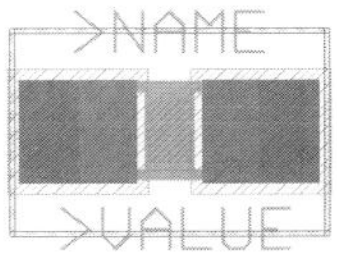 >NAME >VALUE
100 mW	0805	*0805-RES*	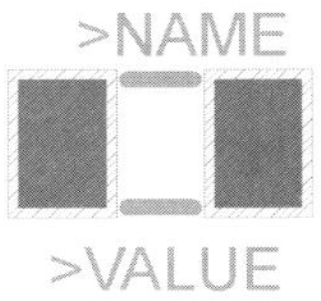 >NAME >VALUE
125 mW	1206	Resistor *1206*	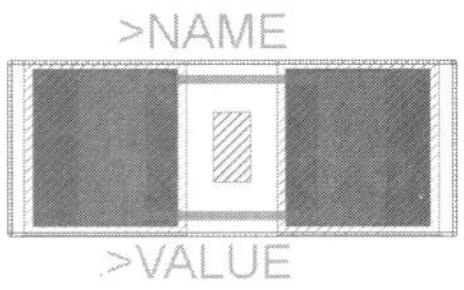 >NAME >VALUE
250 mW	1210	Resistor *1210*	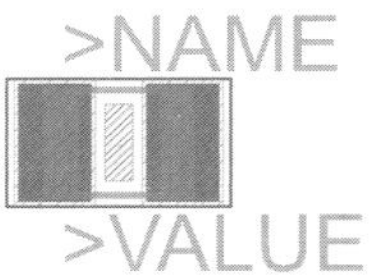 >NAME >VALUE
500 mW	2010	Resistor *2010*	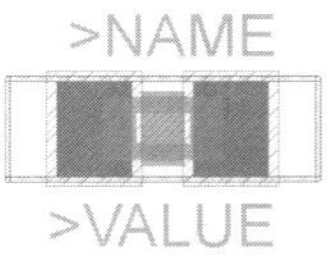 >NAME >VALUE
1 W	2512	Resistor *2512*	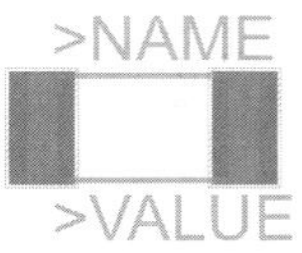 >NAME >VALUE

the resistors in Figure 3-4 have the code 1001. This means three value digits 1, 0, and 0 and a number of zeros digit of 1. Thus the value is 1,000 Ω or 1 kΩ.

Capacitors

If you are developing digital projects that use perhaps a microcontroller and a few extra components, then you will only be using a fairly small set of capacitors and using them in a pretty simple way—probably 100-nF decoupling capacitors close to ICs or perhaps 100-μF capacitors around a voltage regulator.

Through-Hole Capacitors

As decoupling capacitors placed close to ICs, tiny multilayer ceramic capacitors are ideal. These generally will be bead-shaped with leads on a 0.1-in. pitch. Electrolytics are larger and are available with axial leads (emerging from the ends of the tubes). It is more common to use devices with radial leads, where both leads emerge from the same end of the tube. For long, thin electrolytic capacitors, it is a good idea to allow them to bend over and lie flat against the PCB rather than stick up, where they could easily be damaged. Figure 3-5 shows a selection of through-hole capacitors.

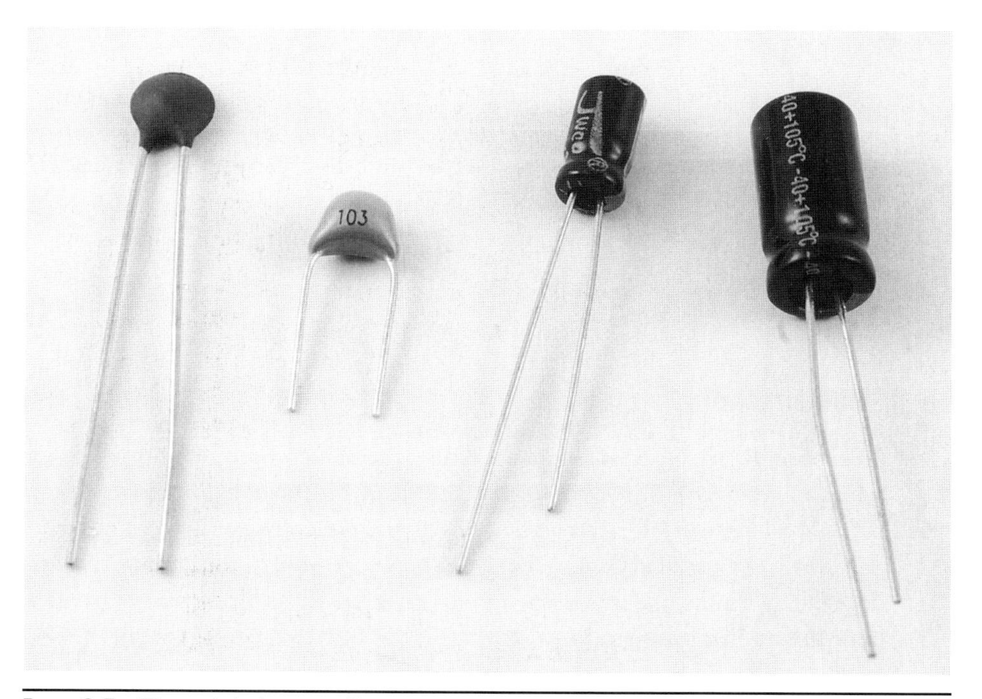

FIGURE 3-5 Through-hole capacitors.

Table 3-3 provides a list of some of the most commonly used through-hole capacitors and indicates where to find them in the EAGLE libraries.

Table 3-3 Common Through-Hole Capacitors

Capacitor	Library	Footprint
100-nF multilayer ceramic and most small capacitors	CAPPTH	>Name >Value
10- to 100-µF electrolytic	CAP_POLPTH2	>Name >Value
100-µF and larger	CAP_POLPTH1	>Name >Value

Electrolytic capacitors usually have their capacitance and maximum voltage written on them. Ceramic and smaller through-hole capacitors are identified by a three-digit code. This is the capacitance in picofarads using a similar scheme to resistor code. The first two digits are the value, and the third digit is the number of zeros to add. Thus a capacitor marked as 104 is 1, 0, 0000, or 100,000 pF or 100 nF.

SMD Capacitors

Small nonpolarized SMD capacitors use the same common footprints as SMD resistors, that is, 0402, 0603, 0805, 1206, and 1210 being common sizes. When it comes to larger-value electrolytic capacitors, typically the devices have a square base with a cylindrical component on top. A commonly used standard for this was devised by Panasonic and is generally referred to as *Panasonic B to E* for the majority of electrolytic sizes.

Figure 3-6 shows a selection of SMD capacitors. The capacitor sizes are (left to right) 0201, 0805, 1206, and Panasonic D. All the capacitors are 100 nF. You

FIGURE 3-6 SMD capacitors.

may at first sight miss the 0201 capacitor. It really is minute. When selecting components for hand soldering, don't attempt anything smaller than 0805.

Table 3-4 shows some of the most common SMD capacitors, their footprints, and where to find them in the EAGLE libraries. Small SMD capacitors are not usually labeled, so try not to mix them up.

TABLE 3-4 Common SMD Capacitors

Capacitor/Inductor	SMD Size	Library	Footprint
100-nF multilayer ceramic and most small capacitors	0603	*0603-CAP*	>NAME >VALUE
100-nF multilayer ceramic and most small capacitors	0805	cap0805	>NAME >VALUE
10- to 100-μF electrolytic	Panasonic C	CAP_POLC	>NAME >VALUE
100-μF and larger	Panasonic D	CAP_POLD	>NAME >VALUE

Transistors and Diodes

Through-hole diodes have similar footprints to resistors. Transistors are a bit more complex, especially power transistors that need to dissipate some heat.

Through-Hole Transistors

By far the most commonly used package for through-hole transistors is the TO-92 package. Higher-power transistors generally will use a TO-220 package, although intermediate-sized packages are occasionally used, as well as occasionally large packages such as the TO-3 for very high-power devices. Figure 3-7 shows a selection of through-hole transistor packages. These are (left to right) TO-92, TO-126, TO-220, TO-264, and TO-3. Table 3-5 shows the two most commonly used packages and their footprints.

Table 3-5 does not include library search terms because you will generally find the exact part in the library, for example, 2N2222, 2N700, or FQP33N10. Always search for the exact part first. If you cannot find the part in the libraries, then try looking on the manufacturer's website and some of the part collections that can be found on the Internet. Generally, unless your part is really unusual, someone, somewhere will have an EAGLE library that includes it. As a last resort, you can always make your own part (see Chapter 11).

Figure 3-7 Through-hole transistor packages.

Table 3-5 Through-Hole Transistor Packages

Transistor Type	Package	Footprint
Small signal and general-purpose transistors (e.g., 2N2222 and 2N7000)	TO-92	>NAME >VALUE
Power transistor	TO-220	A15.2mm >NAME >VALUE I - O

When using TO-220 parts, you will normally get the choice of vertically mounted or lying flat against the board.

SMD Transistors and Diodes

SMD transistors are mostly one of two SMD package sizes, which keeps things simple. Figure 3-8 shows the package types SOD323, SOT23, and SOT223 (from left to right). SOD stands for "small outline diode," and SOT stands for "small outline transistor." With care, all can be soldered by hand.

Table 3-6 shows some devices and their footprints.

Figure 3-8 SMD diode and transistor package types.

Table 3-6 Diode and Transitor SMD Types

Transistor/Diode Type	Package	Footprint
Small signal or rectifier diode, typically up to 200 mA	SOD323	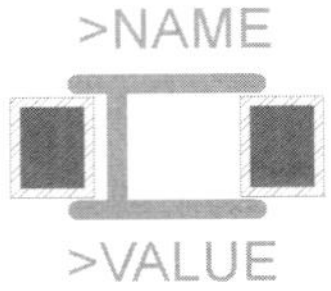 >NAME >VALUE
Small signal and general-purpose transistors	SOT23	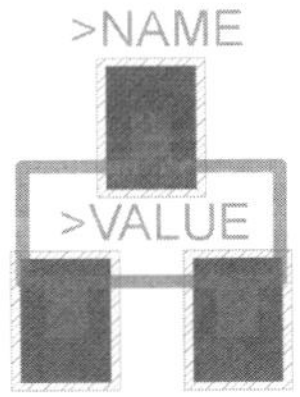 >NAME >VALUE
Power transistor	SOT223	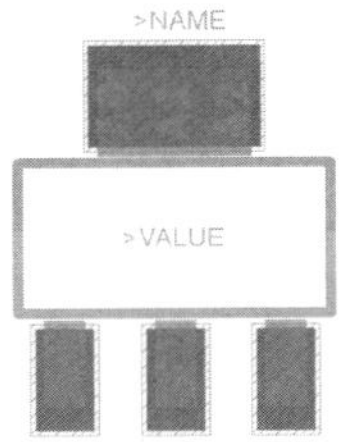 >NAME >VALUE

Integrated Circuits

ICs come in a huge array of different package types. Through-hole devices are much easier to use, generally being dual in-line (DIL) packages, unless they are three-pin devices, in which case they use transistor packages.

On the other hand, there are many standard sizes for SMD ICs. Many are too small to attempt hand soldering, however. We will just cover the devices with pin spacings large enough to solder by hand here.

When finding an IC to add to a schematic, you will often find that the same IC part number is available in a number of packages, both through-hole and SMD (Figure 3-9).

Through-Hole ICs (DIL)

Figure 3-9 shows a couple of DIL IC packages. These can have up to 40 pins, and the most common sizes are 8, 14, 16, 20, 24, 28, and 40 pins. Most have a gap of 0.3 in. between the two rows of pins, but some of the bigger devices have a 0.6-in. gap.

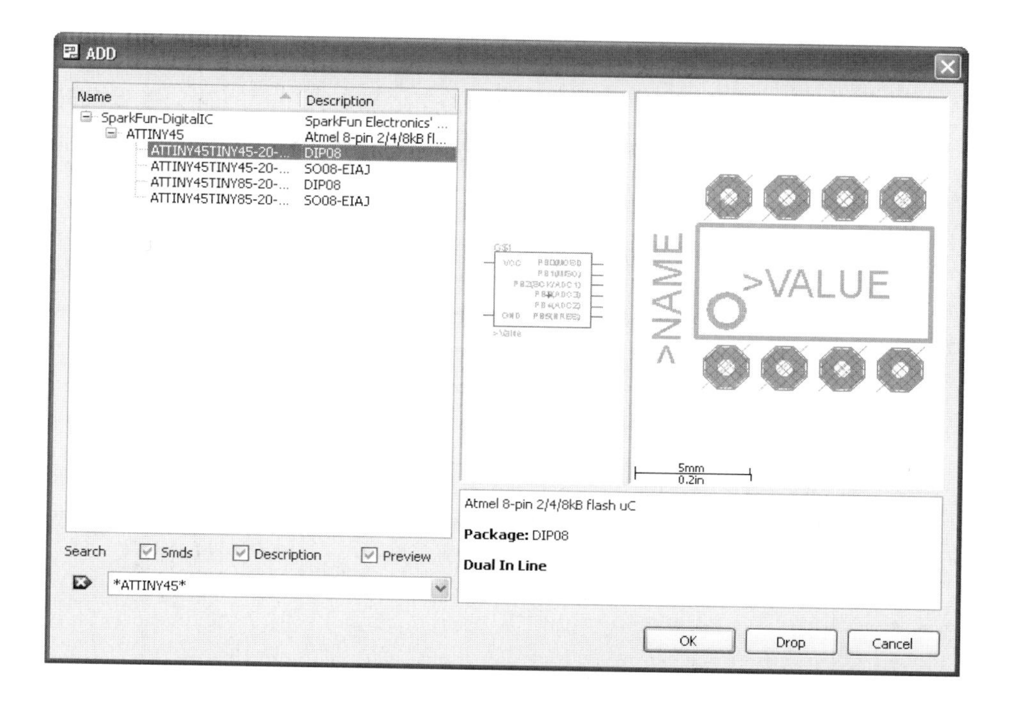

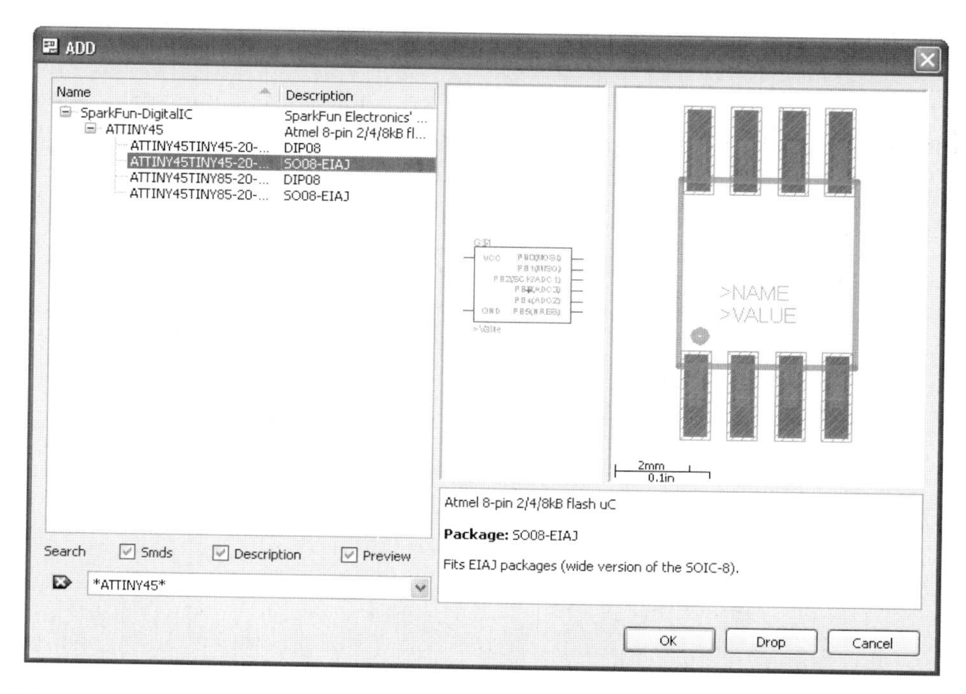

FIGURE 3-9 Package options for an ATTiny45 IC.

SMD ICs

There are a few major styles of SMD IC packages, as summarized in Table 3-7. All these packages have leads protruding from the side of the IC.

Table 3-7 SMD IC Package Types

Package	Connector Pitch	Notes
SO/SOIC/SOP	50 mils (1.27 mm)	Small-outline package
MSOP/VSOP	25 mils (0.65 mm)	Very small-outline package
TSOP	19.7 mils (0.5 mm)	Thin small-outline package
TQFP	0.8 mm	Thin quad flat package

Other package types do not have leads in the conventional sense but rather pads underneath that match up with pads on the PCB. These packages are only really suitable for soldering using solder paste and an oven.

If you expect to be soldering your boards by hand, then you probably should stick to the SO package type. In Chapter 7 you will find some techniques that with practice will allow you to solder some really small pitch devices.

Connectors

Any board that you design is likely to have connectors of some sort. These may be pin headers, JST (Japanese Standard Terminal) sockets, screw terminals, or just solder pads to which wires will be soldered. Figure 3-10 shows a selection of different connector types.

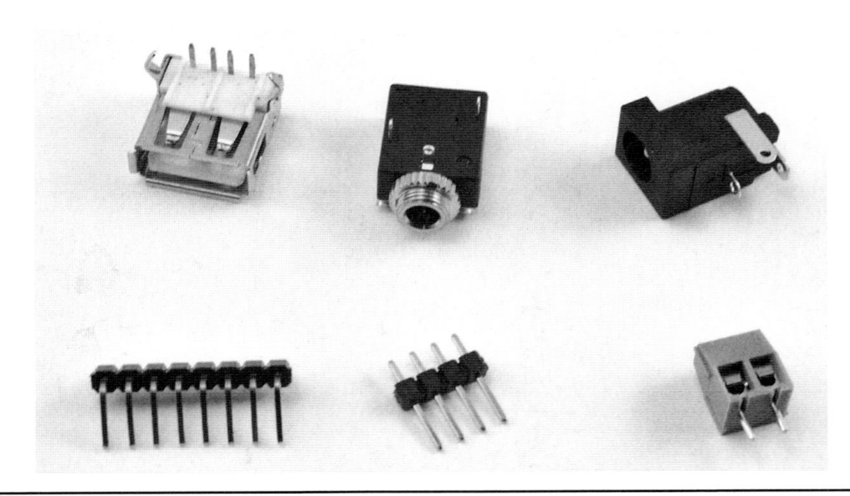

Figure 3-10 Connectors.

Table 3-8 lists some of the more common connectors and where they can be found in the EAGLE libraries. You should find most of what you need in the Sparkfun Connectors library.

Table 3-8 Common Connectors

Connector	Library	Footprint	Notes
0.1-in. pin header	M02PTH3	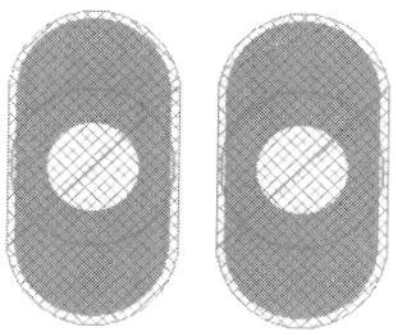	Suitable for data and power connections; low cost
2.1-mm barrel jack	power_jackpth	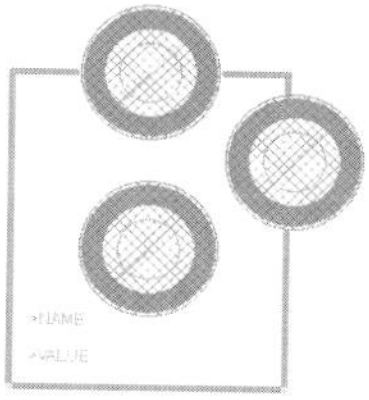	DC power connections
Screw terminals	M023.5MM	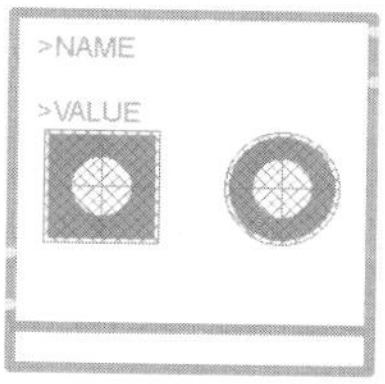	Power, speaker connections, etc.
3.5-mm stereo audio socket	audio-jack2.5mm	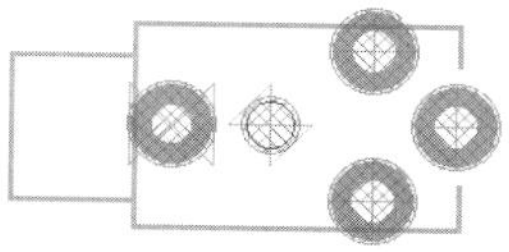	Audio connections
Two-way JST socket	jst	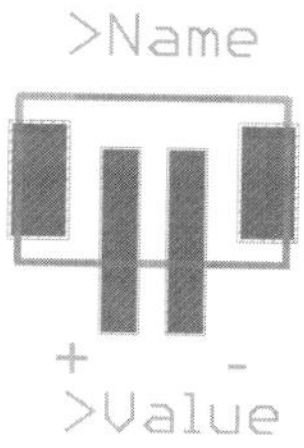	LiPo battery connections

For best structural rigidity, many connectors are through-hole so that they use the board to provide extra strength. Connectors may have mechanical pins that go through the board to provide extra strength. Surface-mount connectors are essentially only attached to the copper track on the surface of the board. This means that under stress, it is easy for the copper layer to lift away from the board.

Other Components

The preceding sections detail the most common components. For other components, it just becomes a matter of finding the part in the library. If there are external components that are to be connected by wires to the PCB, then select one of the connector types and solder the wires directly to the pads.

Buying Components

It may sound obvious, but always make sure that you can easily get hold of a part before you design something that uses it. If you are designing something that will become a product, then keeping the cost of the bill of materials (BOM) to a minimum will become important.

Always use common components except where there is a good reason to use something more exotic. Component suppliers that are particularly suited to the hobbyist include

- Sparkfun: www.sparkfun.com
- Adafruit: www.adafruit.com
- Maplins (UK): www.maplins.com
- RadioShack: www.radioshack.com

None of these has an exhaustive range of components, and they are not particularly cheap, but they combine component sales with modules and very good reference materials and tutorials.

Perhaps the next tier of supplier has component suppliers that have a much wider range, are happy to supply in small quantities, and do not always have a minimum order value. These include

- Mouser: www.mouser.com
- Digikey: www.digikey.com

The top tier of component suppliers can supply almost any part that is in production. They supply mostly to the electronics industry, and their prices often can be surprisingly competitive on some lines, especially if you are buying in quantity. The most prominent of these are

- Farnell (worldwide): www.farnell.com
- Newark (United States): www.newark.com (owned by Farnell)
- CPC (United Kingdom): cpc.farnell.com
- RadioSpares (worldwide): www.rs-components.com

A great web resource for tracking down parts is Octopart (www.octopart.com). This is a component search engine. You just type in the part name, and it will give you a list of suppliers selling the part and their prices.

Often the cheapest place to get components is using eBay. Large quantities of components such as LEDs can be bought directly from China for a very low price.

Paper PCB

It is always difficult to know that the component you have in your hand will exactly fit a particular footprint. This is particularly the case for connectors and unusual components, especially if the components have been scavenged or found in your personal stock of components.

A good way to check that you have the right component is to print out a *paper PCB* from EAGLE so that you can hold the components against the paper to check the footprints. It is not much effort to go to and much better than having made a batch of useless PCBs.

Summary

Now that you can find the components you need for your projects, we can return to the process of drawing the schematic. This time we will take a more thorough approach and look at some of the features of the Schematic Editor that we did not need to use in Chapter 2.

CHAPTER 4

Editing Schematics

In Chapter 2, we very quickly built the schematic for a simple LED flasher using a 555 timer. In this chapter, after some general explanation of the Schematic Editor, we will start developing a second project, a sound-level meter that includes a small amplifier and 10 LEDs to indicate the level of sound. This project will be continued in Chapter 5 when we come to look at laying out PCBs in more detail.

Because you are going to be experimenting with various editor commands to see how things work, it is probably a good idea to create a new "play" project to which you can add a new schematic for experimenting with.

The Anatomy of the Schematic Editor

Figure 4-1 shows a new "Schematic Editor" window. We will now look at the parts of this window in detail and look at what everything does. We will start by identifying and naming the main areas of the window.

Starting at the top of the screen, we have the Title bar. This shows the file name of the schematic being worked on. This can be useful as confirmation that you are actually working on the document that you think you are working on.

Next, we have the Menu bar. This contains options for saving the document, as well as providing an alternative means of accessing many of the functions found on other icon buttons on the editor.

The Action toolbar contains a number of commands and features that are also accessible on the menus. From left to right in Figure 4-1, the icons are

- **Open.** Another schematic.
- **Save.** The current schematic.

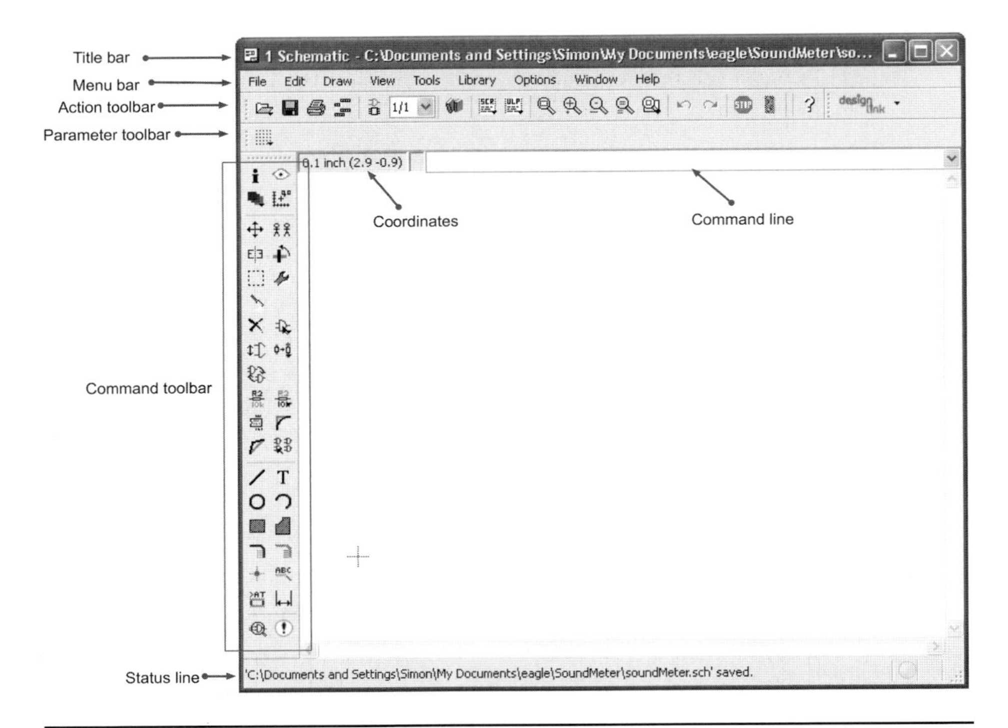

FIGURE 4-1 The "Schematic Editor" window.

- **Print.** Open the print dialog.
- **CAM.** Launch the CAM (computer-aided manufacturing) processor for producing design files suitable for sending to a PCB fabrication shop (see Chapter 6).
- **Board.** Switch to or open the corresponding board design for the schematic.
- **Sheet Drop Down.** We are only allowed one sheet with the light version of EAGLE, so we can ignore this.
- **Use Library.** Add a library to the list of libraries in use on this schematic.
- **Execute a Script.** See Chapter 10.
- **Execute a User-Language Program.** See Chapter 10.
- **Zoom to Fit.** Zooms to just fit all the components in the schematic into the editor area (useful).
- **Zoom In.** It is easier to use the scroll wheel on your mouse if you have one.
- **Zoom Out.**
- **Redraw Screen.** Occasionally, screen redraw is glitchy. This forces a redraw of the screen.
- **Zoom to Selection.**
- **Undo.** Undo the last change made to the schematic (very useful).

- **Redo.**
- **Cancel Command.** Cancel a command that is running.
- **Help.** Search for help about using EAGLE.

Under the Action toolbar, you will find the Parameter toolbar. The contents of this toolbar will change as you select different commands in the Command toolbar. It is the Command toolbar that you will use most when you are editing a schematic. We will deal with this specifically in the next section.

Underneath the Parameter toolbar is a gray area that displays the current coordinates of the cursor. To the right of this is a text entry field called the *command line*, where you can type commands equivalent to pressing buttons. We will look at this briefly in Chapter 9 because generally users prefer to click on commands rather than type them.

The Command Toolbar

The icons in the Command toolbar are labeled in Figure 4-2. There are a lot of commands that we can use. The ones labeled in bold in the figure are the most

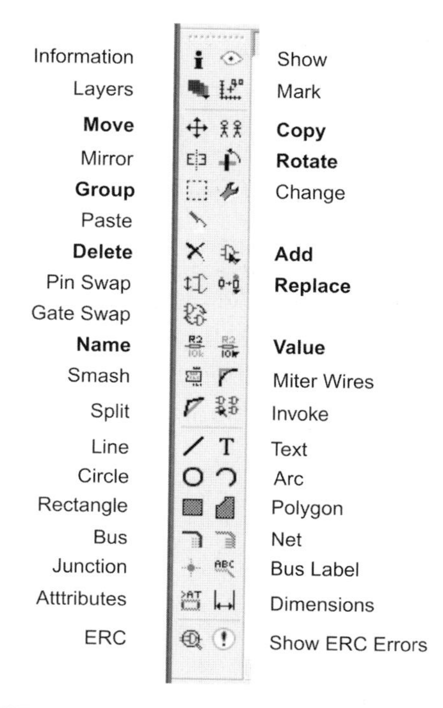

FIGURE 4-2 The Schematic Editor Command toolbar.

commonly used ones. You will use these commands a lot. Some of the others deserve a mention so that you know they are there when you do come to need them.

Common Commands

We have already used some of the common commands in Chapter 3, but we will recap on them here.

Move

With this command selected, you can drag any components or nets around on the Schematic Editor. Sometimes it can be difficult to select the component you want. If you look carefully, you will see that each component has a little cross on it. This is the selection point, so aim for that.

To move more than one thing at a time, you have to use the Group command.

Group

The Group command can be very baffling when you first come to use it. It allows you to select a set of parts by dragging over the set with the mouse. However, to then do anything meaningful with the selected set, you have to think ahead and select a command before you select the Group command. For example, we will look at dragging a group of items from one area of the Schematic Editor to another. This is something you are likely to do fairly often.

The sequence of actions is

1. Select the Move tool.
2. Select the Group tool.
3. Drag the cursor over the items that you want to include in the group move. They will become highlighted.
4. Right-click, and down near the bottom of the Context menu, you will see an option "Move: Group" (Figure 4-3).

Delete

Delete anything you subsequently click on without further warning. You can also carry out this command as a Group command.

Name and Value

These two commands work quite well in the sort of "verb-noun" workflow of EAGLE. Having dropped a load of resistors onto the schematic, you will probably want to go around them setting their names and values.

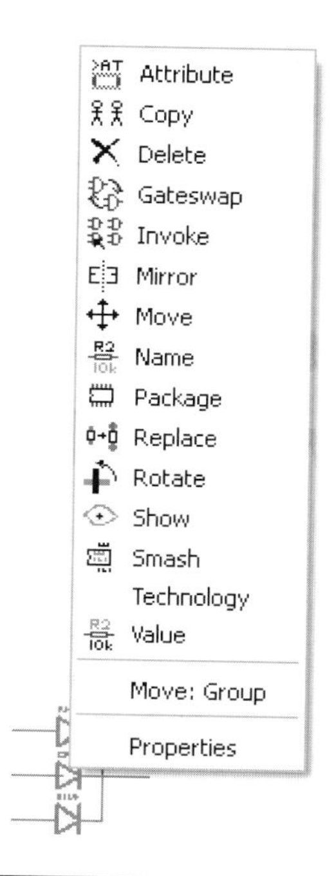

FIGURE 4-3 "Move: Group" on the Context menu.

Copy

Copy is a useful command that we have not met before. It should perhaps more accurately be called "Clone" or "Duplicate."

To use it, select the "Copy" icon, and then click on the item you want to make a copy of. The copy will become attached to the cursor, which you can then move and click to deposit the copy onto the schematic.

Rotate

To rotate a component, click on the Rotate command, and then click on the component to be rotated. This will rotate it through 90 degrees. To rotate it more, just keep clicking.

When drawing a schematic, it only really makes sense to rotate through multiples of 90 degrees. However, when designing the board layout, it is occasionally useful to be able to rotate a component to an unusual angle such as 45 degrees.

TIP *When you are moving components around, you will often find that you need to switch between moving and rotating. You can rotate a component when using the Move command by right-clicking.*

Add

We have dealt pretty well with Add. Once you have selected the component that you want to add, then every click of the mouse will result in a new component being dropped onto the design area. To go back and select a different component to add, click on the ESC key.

One aspect of the Add command that we have not looked at is adding multigate or multipart components. Examples of this kind of component include logic chips that contain multiple gates or chips such as the NE556 that contain the equivalent to two NE555 timers in a single chip.

When you add one of these components, the first click will put the first gate or subpart onto the design area; subsequent clicks will place the other gates or subparts. Each part will be given a name, the first part of which is the component number and the second part of which is the gate or subpart within that chip.

Replace

Replace is a useful command that lets you swap out one component for another that is pin-compatible. This has the advantage over deleting it and adding a new part in that it retains all the net attachments to the part.

To use it, select the command, and then browse for the replacement component. Then click on the component in the design area that you wish to replace.

Net

We will cover nets in a later section.

Other Commands

Other commands that you will need from time to time are detailed below. It is worth gaining some basic familiarity with them.

Information

With this command selected, every time you click on an item in the editor, you will see an information window like Figure 4-4 that shows you information about that item.

This actually provides an alternative way of changing information about the item, such as its name and value.

Layers

You can control what information is visible on the schematic. This is not really very useful on a schematic but becomes essential when we look at board layout in Chapter 5.

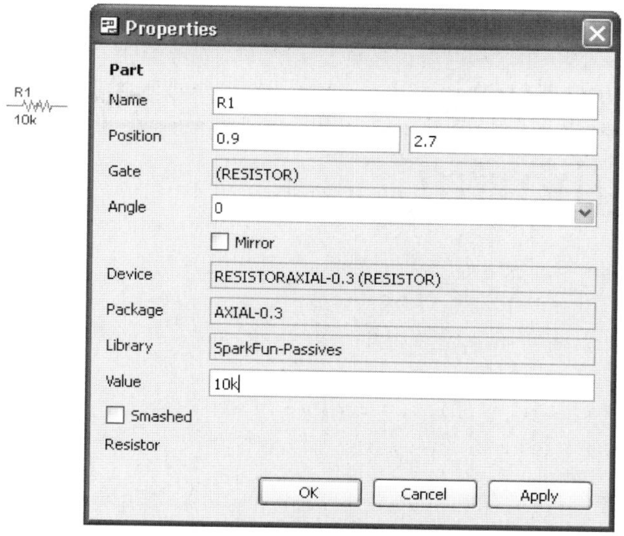

FIGURE 4-4 Information about a resistor.

Mirror

Mirroring a component flips it through its vertical axis. Figure 4-5 shows a transistor and a mirrored transistor next to each other. It just allows the circuit symbol to be displayed the other way around.

Paste

Paste is the same concept as the Paste in Cut and Paste, but this being EAGLE, it works in a different way. To make a copy to be pasted, you select items either individually or using the Group command. The Paste command will then attach a duplicate of everything selected to the cursor so that you can click to place it in the Schematic Editor.

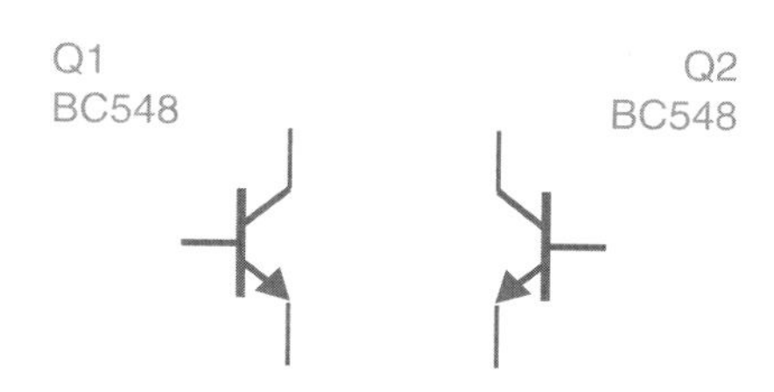

FIGURE 4-5 A mirrored transistor.

Pin Swap

Some components such as logic gates have pins that are interchangeable. The Pin Swap command allows you to swap the pins. There is no reason why you would want to do this while designing a schematic, but when routing the PCB, you might find that things would become a lot easier if you could just swap over which pin is used.

Gate Swap

As with Pin Swap, this feature only becomes useful when you start designing the PCB layout and suddenly discover that the layout would be much easier if only you had used a different gate or subpart of the chip. By switching back to the Schematic Editor, you can fix this.

Smash

This rather dramatic sounding command is nowhere near as exciting as you might expect. It simply allows you to separate the Name and Value labels that accompany a circuit symbol from the symbol itself so that you can move them around independently. This helps to keep your schematics clear and easy to read.

Split

The Split command allows you to add a waypoint to a net line. Select the command, and then click on the length of the line somewhere. You can then drag a point out, and the line will follow. This is useful for changing the path of a net line without having to delete it and redraw it. Note that you can also use the Move command on net lines.

Line, Circle, Rectangle, Text, Arc, and Polygon

All these commands allow you to add decorations to a schematic. These take no part in the electrical side of the schematic. They will not have any influence over the board layout; they just allow you to add further information to the schematic diagram.

Bus and Bus Label

Schematic diagrams can become messy, especially in digital electronic designs when you have a lot of wires running from one chip to another. To keep things neat, the Bus command allows you to group the net lines together. See the separate section on buses later in this chapter for more information.

Junction

EAGLE does a pretty good job of automatically marking junctions between one net and another with a little blob. The Junction command allows you manual control over this process. You may never use this feature, though.

Attributes

You can add your own custom attributes to a component and then decide if you want them to be displayed. Figure 4-6 shows a POWER attribute being added to resistor R5. We have also specified that just the value (1 W) should be visible.

If the attribute is displayed in the wrong place, for example, overlapping the component graphic, then you can move it by using the Smash command to separate the circuit symbol from its labels.

ERC and Show ERC Errors

These two commands launch the electrical rule checker (ERC) and open the window showing the results of the check, respectively. We touched on this in Chapter 2 and will meet it again later in this chapter.

Show

The Show command has a similar purpose to the Information command, except that rather than opening a little window giving you details of the component selected, it displays the information in the status area at the bottom of the screen

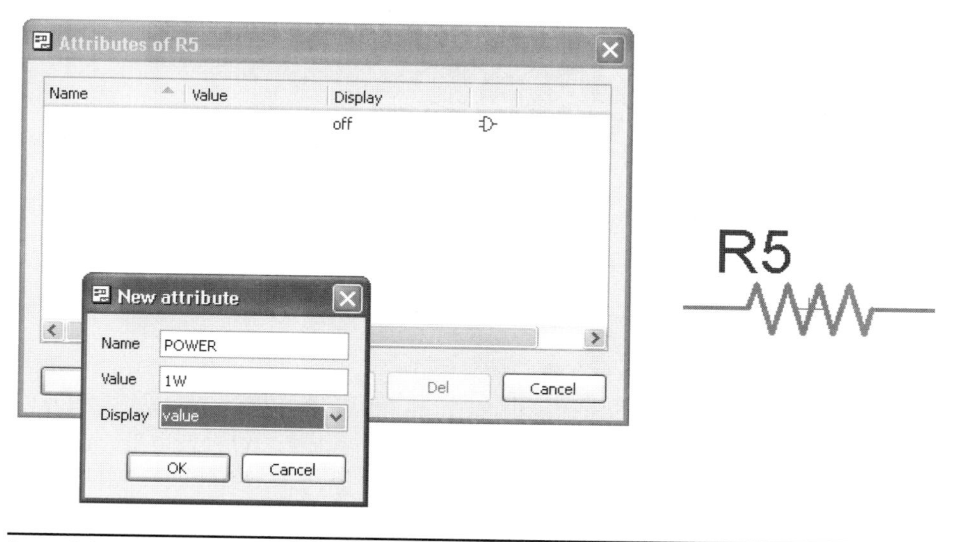

FIGURE 4-6 Adding an attribute.

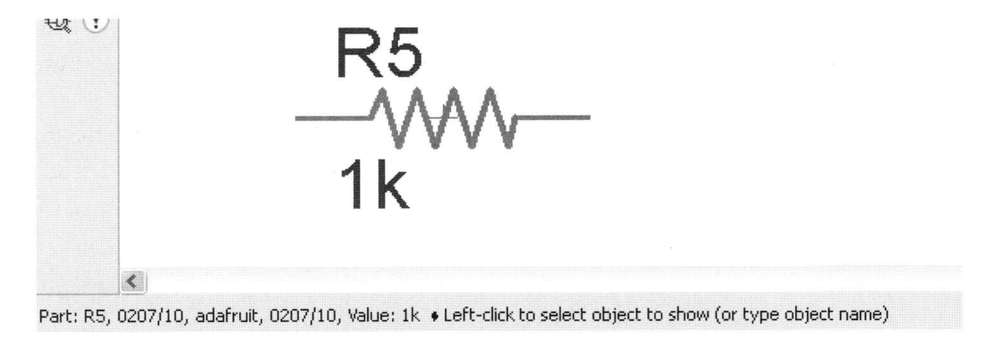

Figure 4-7 The Show command.

(Figure 4-7). Because it is not opening a new window (that then needs closing), this can be a faster way to see what's on a schematic.

Mark

This is not something you are ever likely to use on a schematic. The Mark command allows you to set a local origin anywhere on the schematic so that you can see the coordinate values relative to that point. Clicking on the "GO" button (looks like a traffic signal) cancels the Mark command.

Change

The Change command allows you to change almost anything about an object on the schematic diagram. You can, of course, change these things using the Information command. The Change command is not that useful on a schematic but comes into its own when you are designing a board layout because it lets you easily change such things as track widths, silk-screen fonts, and so on.

Miter Wires

You may never use this feature when editing a schematic. It allows you to put a curve in your net lines. To use it, select the command, and then click on a corner of a net line. You can then set the radius in the dropdown in the Parameter toolbar.

Invoke

This is possibly the worst-named command in EAGLE. You might think that it runs a script. No, not at all. This command would perhaps be better named Fetch. It is used in multipart components such as logic gates. As you saw in the section on the Add command, when adding such components, they appear a gate at a time each time you click on the design area.

The Invoke command gives you more control over this process, allowing you to select the order in which the components are added. This serves little practical purpose.

However, the useful thing you can do with Invoke is to add the power connectors for an integrated circuit (IC). After clicking out four gates from, say, a 7400 IC, if you select the Invoke command and click on one of the gates, a menu like the one shown in Figure 4-8 will appear, allowing you to select either the positive supply for the chip or the negative supply. To add both, action the command twice.

These pins then can be explicitly connected to the appropriate power nets. Note that ICs such as this will have specific net names associated with the power pins, so you do not have to connect them up explicitly as long as you pay attention to the ERC results to make sure that the power nets are all defined correctly.

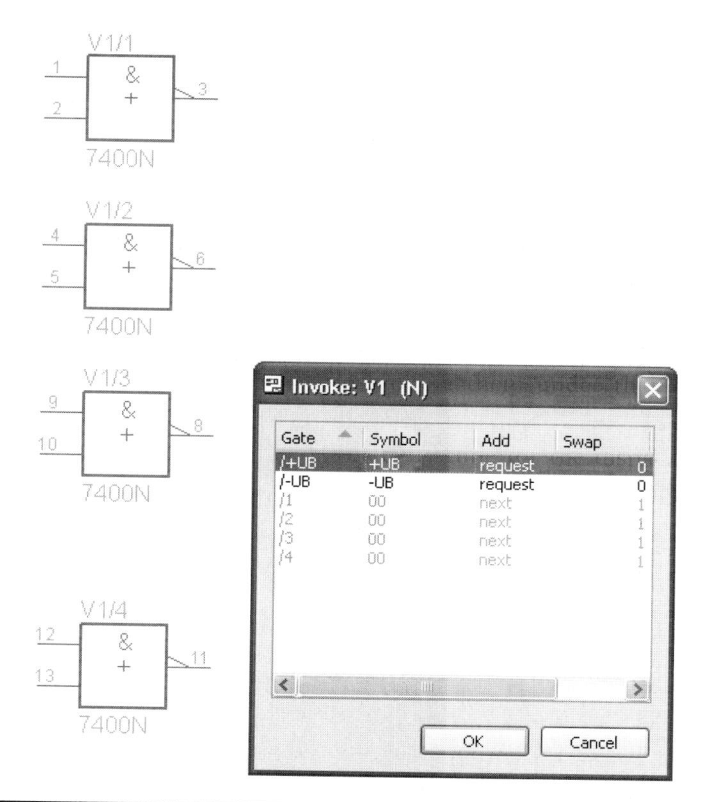

Figure 4-8 Using Invoke to add power-supply pins.

Dimension

This command adds dimension labels. There is no reason to do this in a schematic diagram.

Nets

Aside from the components themselves, nets are the most important part of any design. When you are drawing a schematic, you will use nets to connect one component lead to another. This may be done directly or by connecting a net from one lead of a component to a net that already exists on the diagram.

Let's create a few nets, see how they get named, and see what happens when we connect one net to another. Start by adding four resistors and a net between R1 and R2 and also between R3 and R4 (Figure 4-9).

When adding nets, you can click at any point to add a waypoint (bend in the line), although for this exercise the resistors are lined up so that you don't need any corners. When you get to the end point for a net such as the left-hand lead of R2, the net should stop drawing. It is a sure thing that you missed the connection point on the component if the net unexpectedly carries on drawing after you think you have connected it.

If you click the Show command and then select the net between R1 and R2, it will probably tell you that the net is called N$1, and the other net will be called N$2.

Now draw another net between the two nets, starting from the middle of the net between R1 and R2. When the net reaches the net between R3 and R4, a little window will pop up (Figure 4-10).

This is offering us a choice of names for the new net that will be the merger of the two nets. It does not matter what name you chose.

FIGURE 4-9 Adding nets.

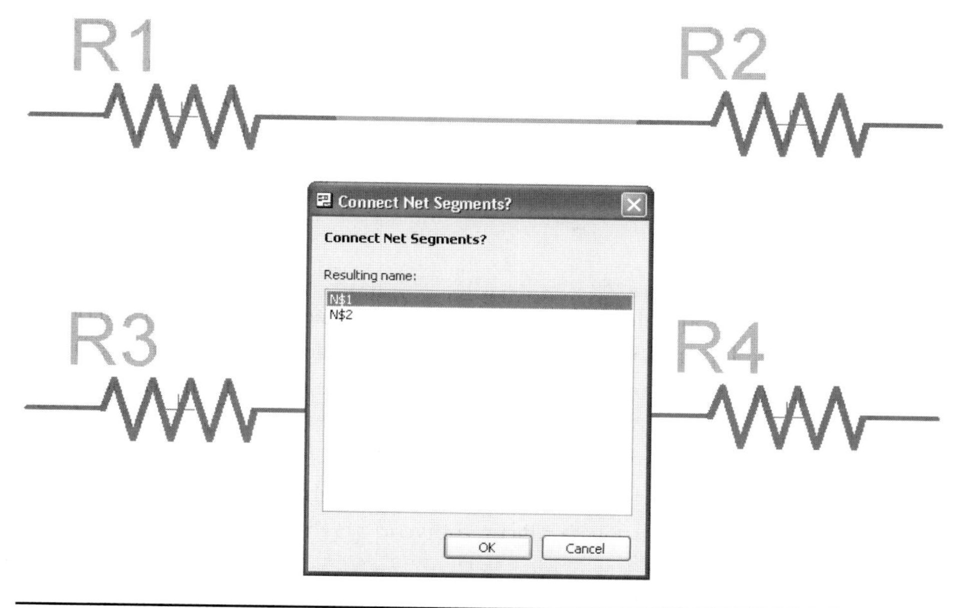

Figure 4-10 Connecting two nets.

Buses

A typical example of where a bus is useful would be where a microcontroller is connected to an LCD module using the HD44780 chip. This requires four data lines and three control lines. In the following example, we will use a bus to connect the display's data and control pins to the microcontroller.

Add the two components (search the libraries for "atmega32u2" and "hd44780"), rotate them, and add a bus, all as shown in Figure 4-11.

Now here's the trick: to allow connections to the bus, a special convention is used in naming the bus. Change the name of the bus to "LCD_BUS1:DB[4..7], E,RW,RS." You can change the name of the bus using the Name or Information commands. This name tells the bus object that any net being connected to it can be connected to one of eight possible slots: DB4, DB5, DB6, DB7, E, RW, or RS. The square brackets and ".." specify a range of values. We could also have set the name to be "LCD_BUS1:DB4,DB5,DB6,DB7,E,RW,RS," and it would work just as well.

Now, when you drag out a net from, say, PD7 on the microcontroller to the bus, it will present a list of possible connection slots (Figure 4-12).

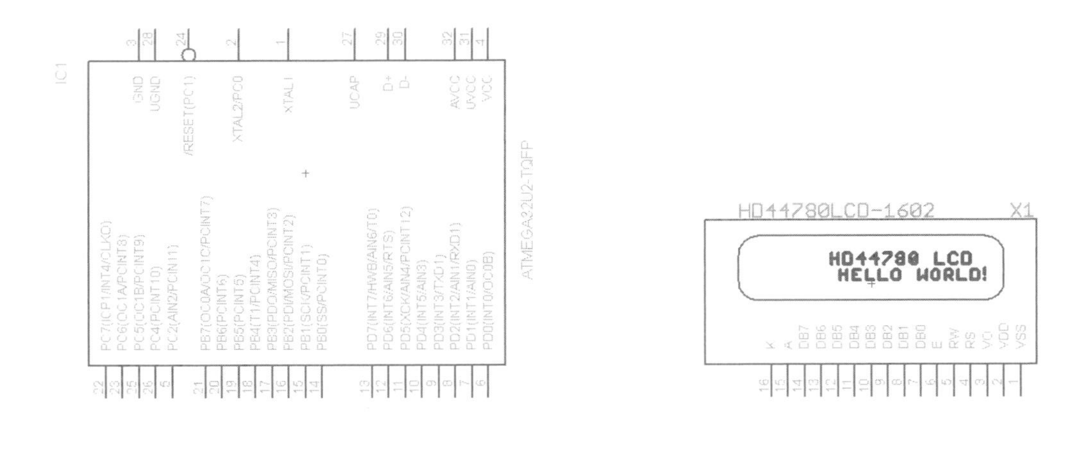

Figure 4-11 Two components and a bus.

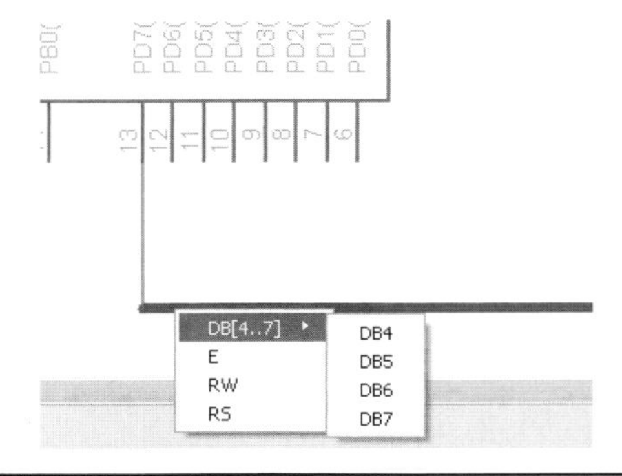

Figure 4-12 Connecting a net to a bus.

At the display end of the bus, you connect nets from the display to the bus using the corresponding slot names.

Worked Example

Now that you are familiar with the tools available to us with the Schematic Editor, we can start building up a second example schematic. The example project is a sound meter. This will use an amplifier chip, a bar-code chip, and a set of LEDs. In Chapter 5, we will design first a through-hole and then a surface-mount version of the PCB layout for the example.

Figure 4-13 shows the finished schematic that we are aiming for. You will notice that it looks considerably more professional than the schematic from Chapter 2.

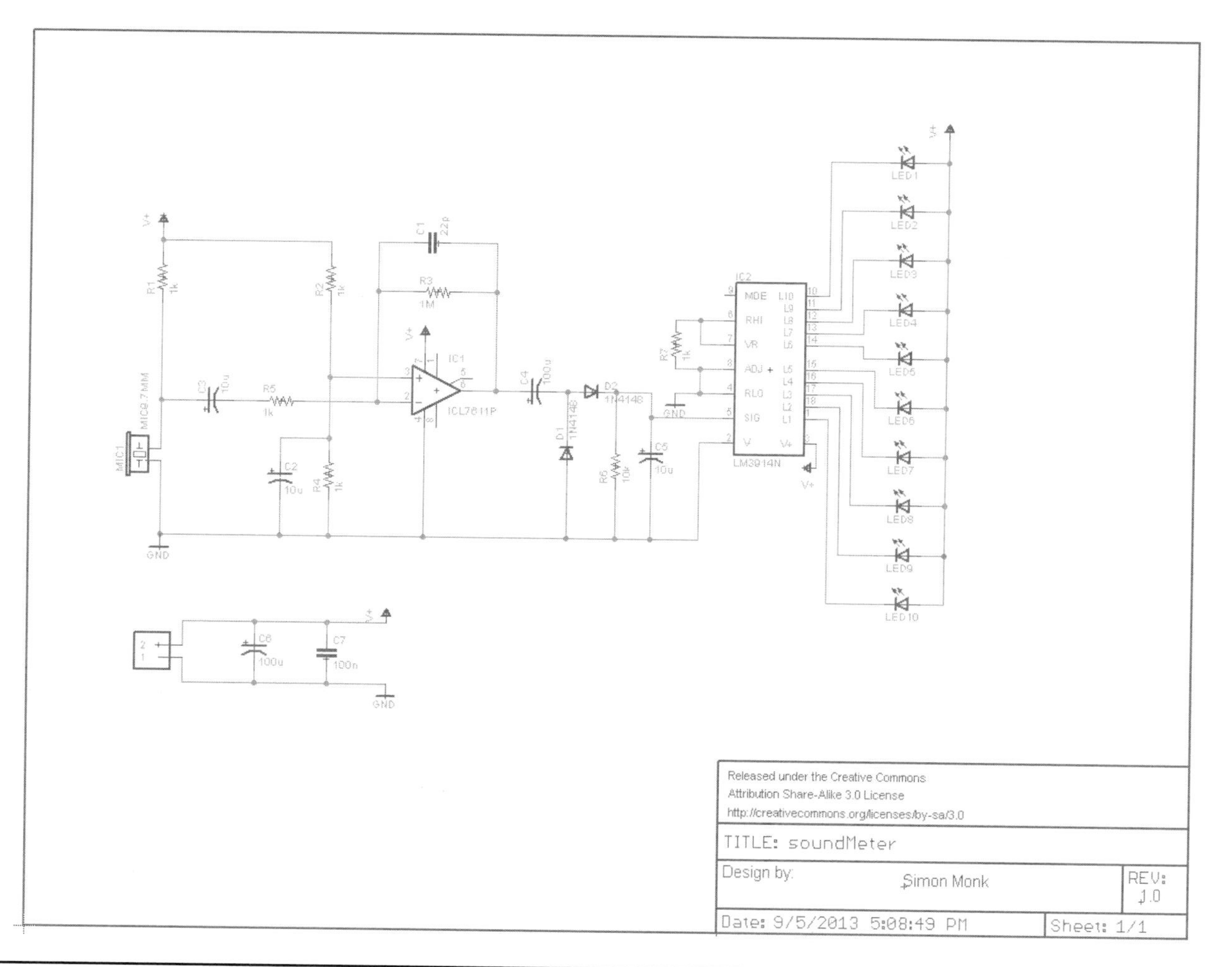

FIGURE 4-13 Final schematic for the bar-code example.

Starting the Schematic

If your design is a one-off, is just for you, and is never likely to be released for others to use, then you can take the approach that we did in Chapter 2. However, you never know where a design might lead, so it can be a good idea to design neatly.

When you look at other people's schematic designs, you will often see the schematic itself framed, with an information panel that provides useful information such as the name of the document, the author, version, and other information such as the licensing of the design. The frame used in Figure 4-13 is the letter-sized frame from the Sparkfun-Aesthetics library and indicates that the design is released under a creative commons license.

The frames can be added as if they were any other part. To add a frame, search the library for "frame." This will bring back frames of different sizes. The size "Letter" will be fine for small designs; for bigger designs, "A3" is a better option. If a design grows unexpectedly, then you can always replace the frame with a bigger one.

Place the frame with the bottom-left corner at the origin. Some of the details of the frame, such as the document name and date last saved, will be automatically shown on the sheet. Note that for them to be updated, you may have to select "Redraw" from the View menu. To change the text in other fields, such as the "Design By" and "Rev" fields, there are two possible ways. The neater way only works if the lettering in the frame has field markers, for example, ">AUTHOR." If it does, then from the Edit menu, select the option "Global Attributes..," and add a new attribute and value, where the attribute name is the same as the word in the frame but without the > character on the front. Then redraw the screen to see the attribute value.

Unfortunately, most of the frames in the library, including the Sparkfun library, do not have such variables defined, so to add the text, use the Text command from the toolbar, enter the text, but before you drop it over the field, change the layer in the Action toolbar to be layer "94 - Symbols." You can also change the text size here or at any time using the Information command, clicking on the text.

Adding the Components

The final set of files for this design, and all the designs in this book, can be downloaded from the author's website (www.simonmonk.org). However, I recommend that you follow the instructions to draw the design for yourself from scratch.

Add the Components

Everyone has different ways of working, but I find it easiest to add all the components that I am going to need, lining them up neatly before moving them to where they are likely to be needed on the schematic.

Add the items in Table 4-1 to the schematic. Where possible, the components are from the Sparkfun library. Searching on the values in the "Device" column works best when searching for the components in the libraries.

Table 4-1 Components Used in the Example Project

Part	Value	Device	Package	Description
C1	22 pF	CAPPTH	CAP-PTH-SMALL	Capacitor
C2, C3, C5	10 µF	CAP_POLPTH2	CPOL-RADIAL-10UF-25V	Capacitor polarized
C4, C6	100 µF	CAP_POLPTH2	CPOL-RADIAL-10UF-25V	Capacitor polarized
C7	100 nF	CAPPTH	CAP-PTH-SMALL	Capacitor
D1, D2	1N4148	1N4148DO35-7	DO35-7	Diode
IC1	ICL7611P	ICL7611P	DIL08	Op amp
IC2	LM3914N	LM3914N	DIL18	Dot/bar display driver
LED1-10	—	LED3MM	LED3MM	LEDs
R1, R2, R4, R5, R7	1 kΩ	RESISTORPTH-1/4W	AXIAL-0.4	Resistor
R3	1 MΩ	RESISTORPTH-1/4W	AXIAL-0.4	Resistor
R6	10 kΩ	RESISTORPTH-1/4W	AXIAL-0.4	Resistor
U$1	MIC9.7MM	MIC9.7MM	MIC-9.7MM	Omnidirectional electret microphone
J1	1X2-3.5MM	1X2-3.5MM (1X2)		Two-way screw terminal

After all these components have been added to the schematic, it will look something like Figure 4-14. Note that this figure does not show the screw terminal J1.

Having added the basic component shapes, you can then go through and set their values according to Table 4-1. D1 and D2 have rather long and overly specific values, so change them to be just 1N4148. You will have to override the warning that they have no user-definable values. This is also a chance to rename the microphone to something more obvious such as MIC1.

With all the components named and the terminal block J1 added, your collection of components should be looking like Figure 4-15.

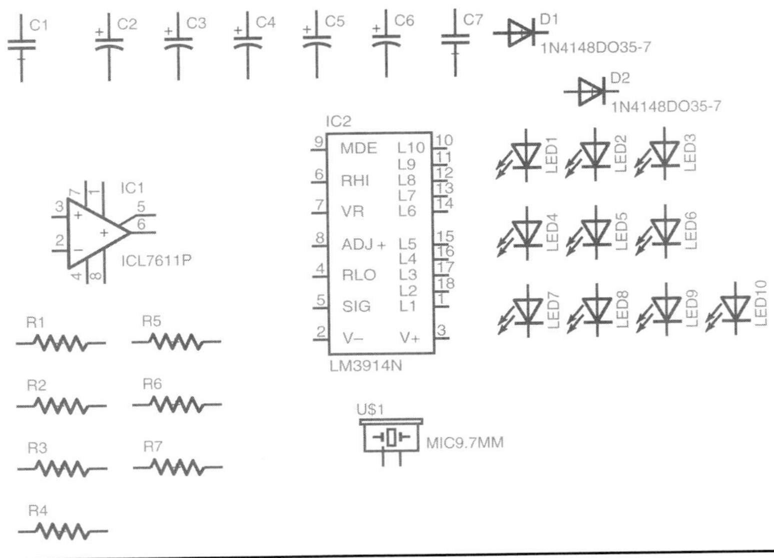

FIGURE 4-14 Schematic with all components added.

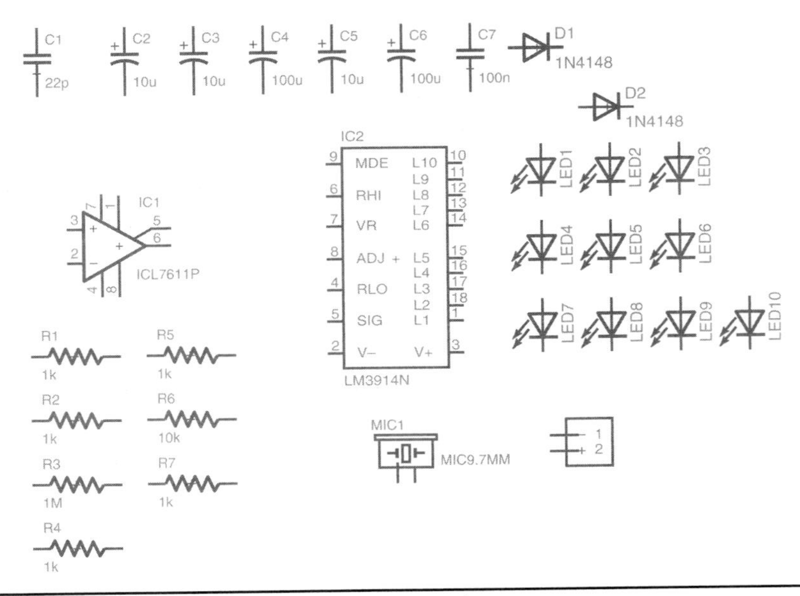

FIGURE 4-15 Setting component values and renaming.

We can now start positioning the components in the approximate location where they will be needed. It is best to place the big multipin devices first, fairly centrally, and then place the other components around them. Logically, the amplifier stage using IC1 happens before the LED bar-code chip does its thing, so put the op amp to the left.

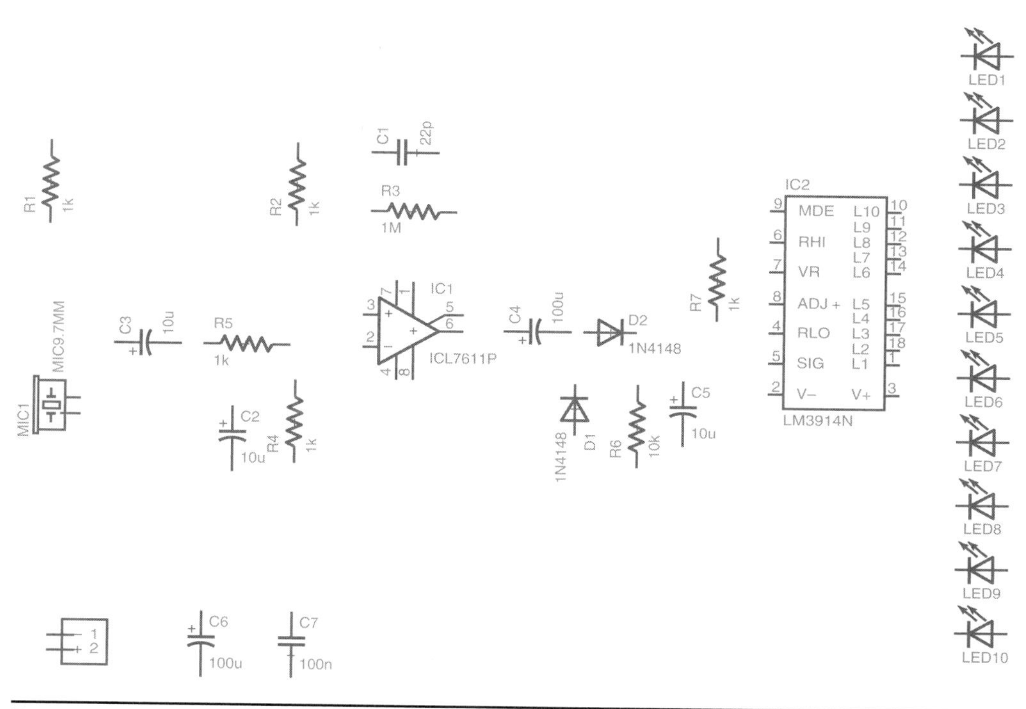

Figure 4-16 The components in the correct positions.

Position and rotate all the components where necessary until they look like Figure 4-16. Remember that you can rotate when you are in Move mode just by right-clicking the mouse.

Adding the Supplies

In our first design, back in Chapter 2, we glossed over the whole idea of supplies. Power came from two pins of a connector, and we did not do anything special to identify them as power rails. We didn't have to. It worked as it was, and that was our rough and ready chapter.

However, generally, except from the very simplest designs, there are good reasons to identify which nets are supply nets and both name them appropriately and associate them with special "supply" parts. This can help to keep diagrams neat because you can repeat GND and other supply symbols so that you do not need to connect every point on the diagram that is GND to every other GND with a line—something that very rapidly gets unmanageable as a design becomes more complex. You can just use the GND symbol as a stand-in.

Therefore, we are going to add two new supply parts, one for GND and one for V+. Select the Add command, and then search for "supply" (Figure 4-17).

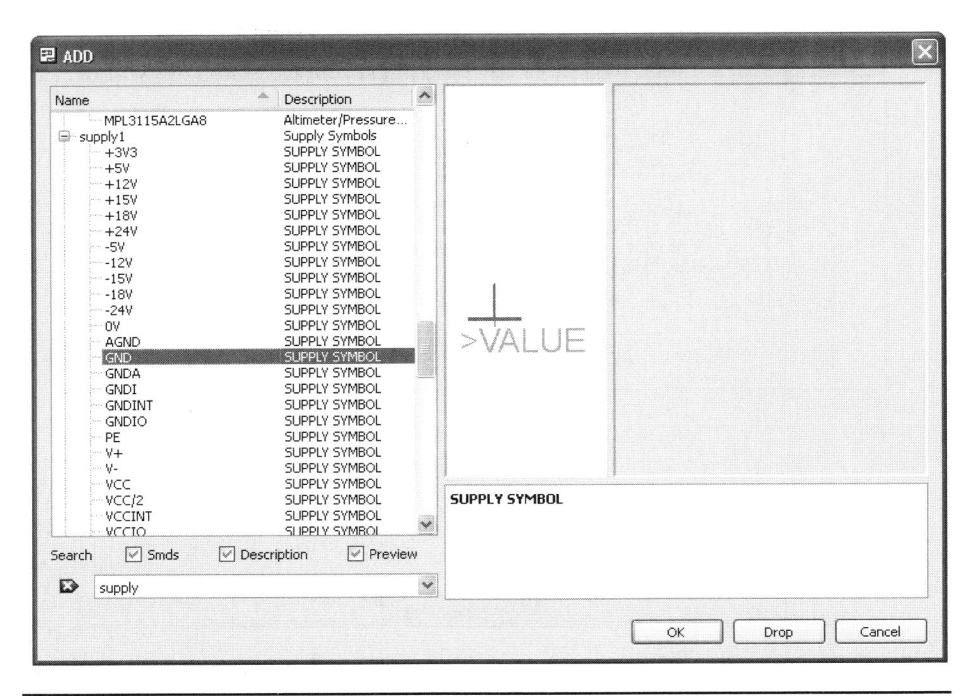

Figure 4-17 Supply parts in the libraries.

As you can see, there are many parts to choose from. We will just use GND and V+ in the `supply1` library. Add one of each near C6 and C7 at the bottom of the diagram and also +V above R1, above pin 7 of IC1, and right next to pin 3 of IC2 and near LED1. Add GND symbols below the Mic and near pin 4 of IC2. When these are in place, your schematic should look like Figure 4-18. Note that in this figure we have also rotated J1 because it was the wrong way around for making connections in Figure 4-16.

You may find that putting all the components in place in one go just gets too complicated, in which case it can be easier to start connecting up some of the nets. However, once you start connecting things together, moving things gets more tricky because you need to worry about where all the lines are going.

Adding the Nets

It's time to start adding the nets, and no doubt, we will need to move things around a little, but hopefully not too much. We will draw in the nets for this example in four stages, corresponding to four areas of the design (Figure 4-19).

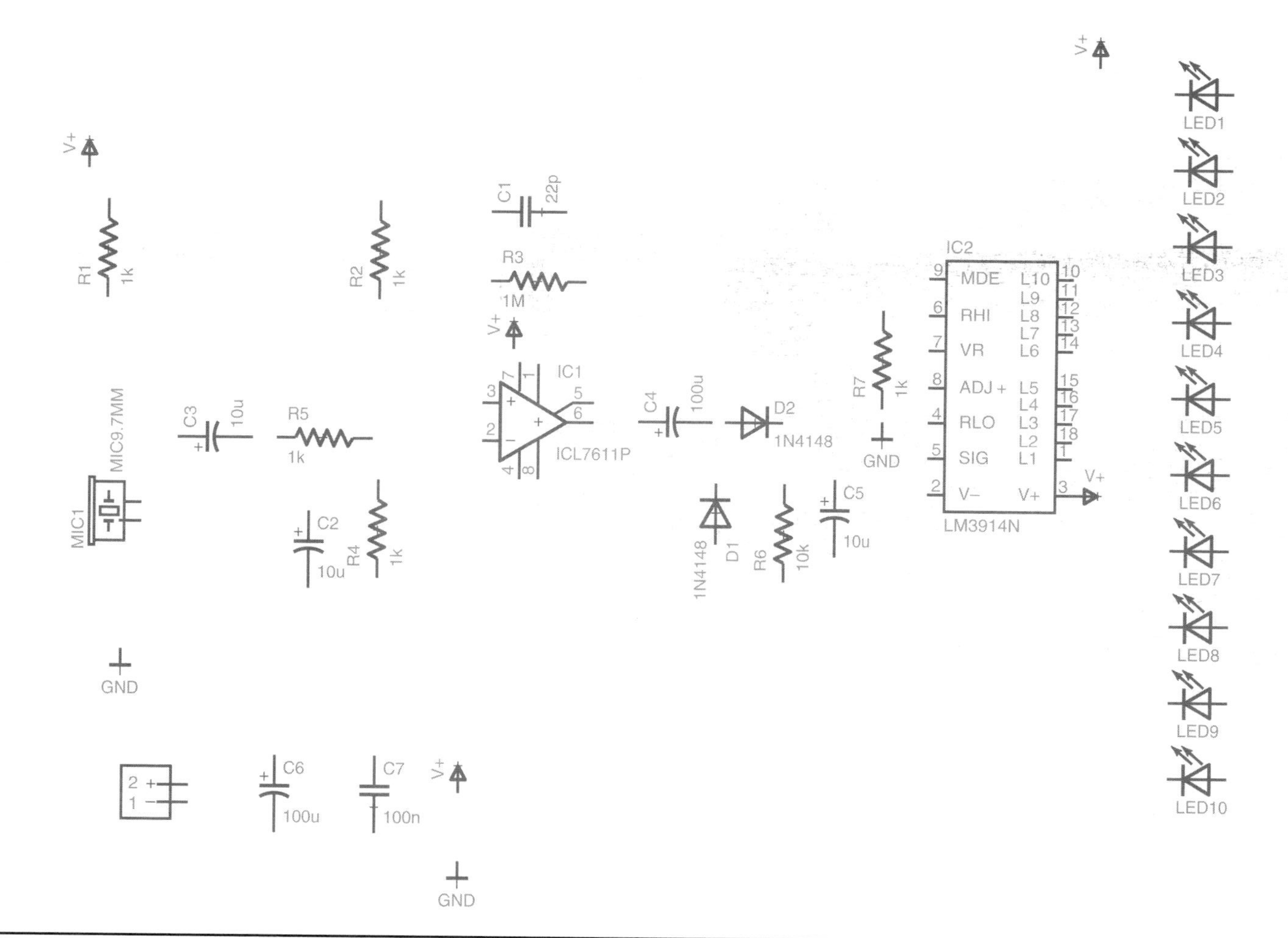

FIGURE 4-18 Supply parts added to the schematic.

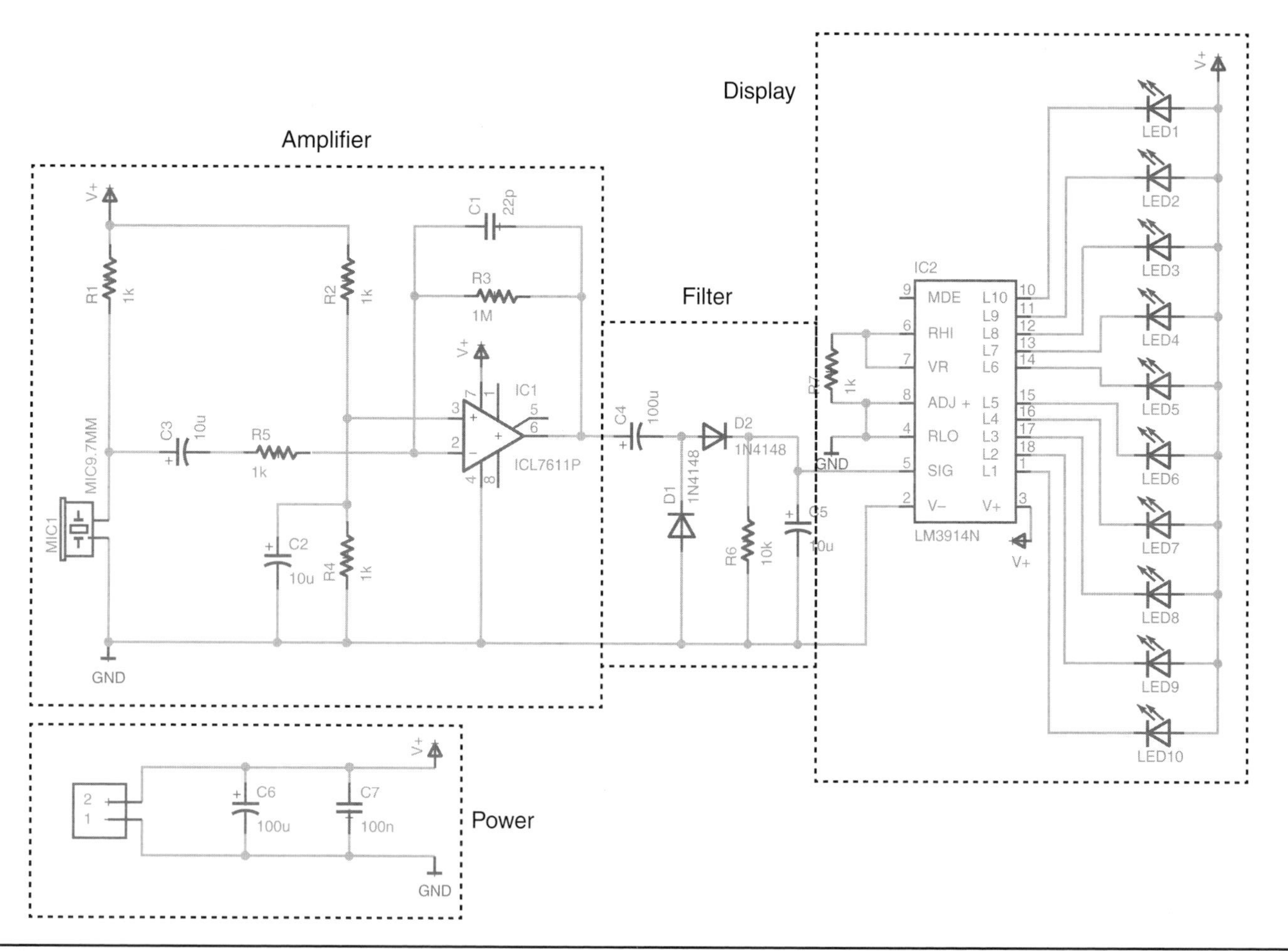

FIGURE 4-19 Logical areas of the design.

The first trivial area is the area at the bottom of the screen comprising the power connector and the two decoupling capacitors. Then we have the amplifier section, centered around IC1, the low pass filter between C4 and C5, and then the bar-code area around IC2.

Let's start by connecting nets between the parts in the power area at the bottom. Figure 4-20 shows the sequence of connections being made.

By starting with connections between the screw terminals J1 and the supply symbols, it makes it easy to then reliably connect up C6 and C7. The temptation to just run the supply line level across the top of the components (Figure 4-21) should be resisted. This will not actually connect them to the supply nets. You can tell that they are not connected because there are no little dots where the capacitor leads meet the GND and V+ nets.

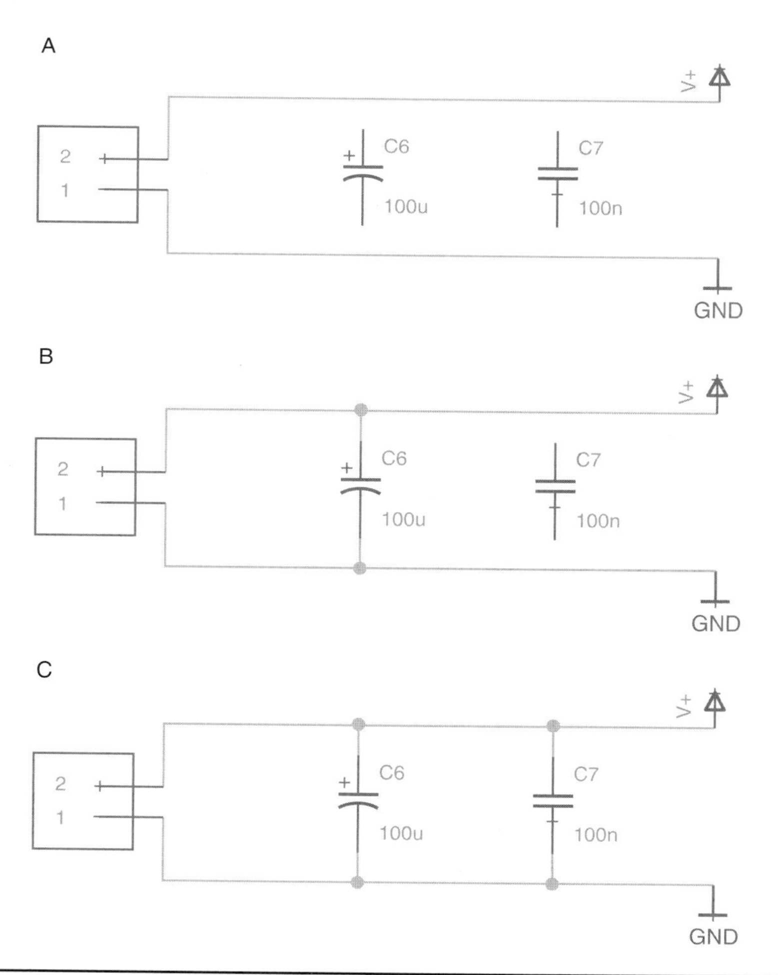

FIGURE 4-20 Adding nets to the supply area.

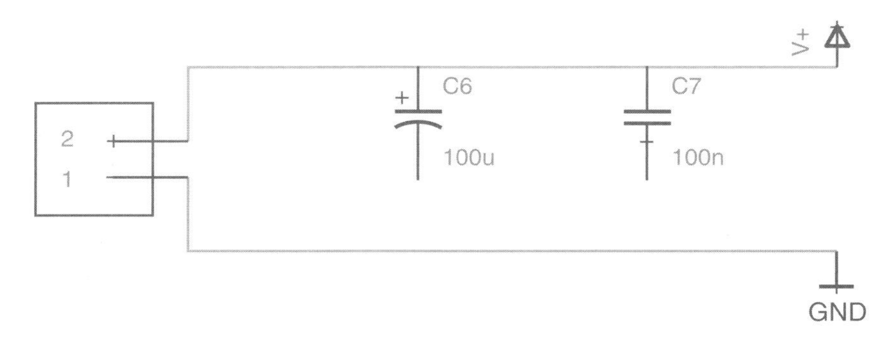

FIGURE 4-21 How not to connect nets.

Now turn your attention to the amplifier stage, around IC1. This is probably the busiest section of the design. Working from left to right, Figure 4-22 shows how I connected these up, starting with the supply connections. Some of the parts will need moving a little so that everything lines up nicely.

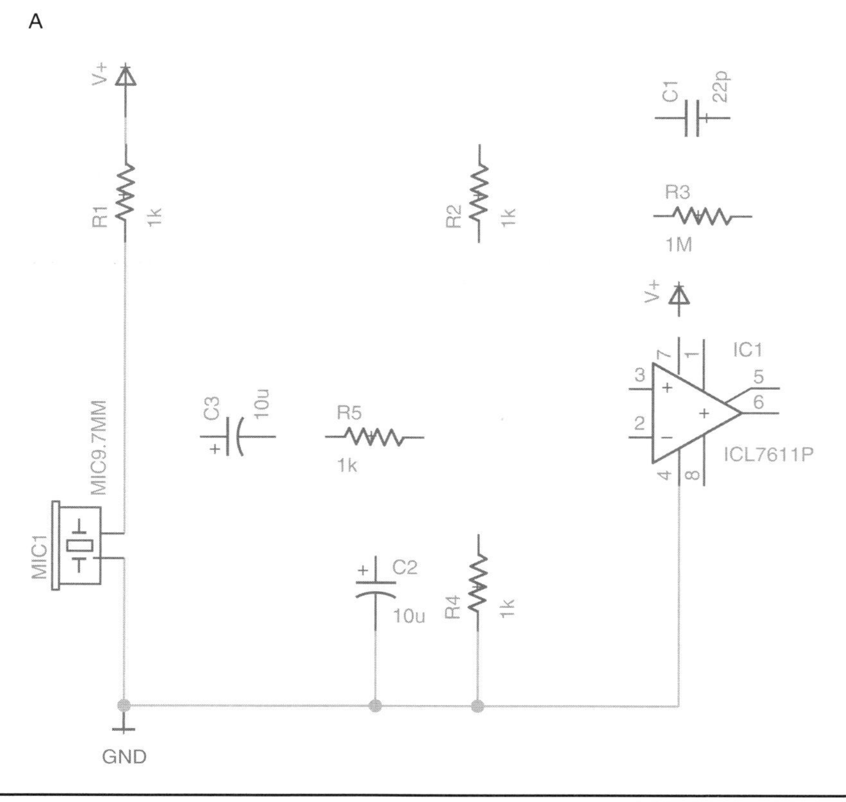

FIGURE 4-22 Connecting up the amplifier section.

B

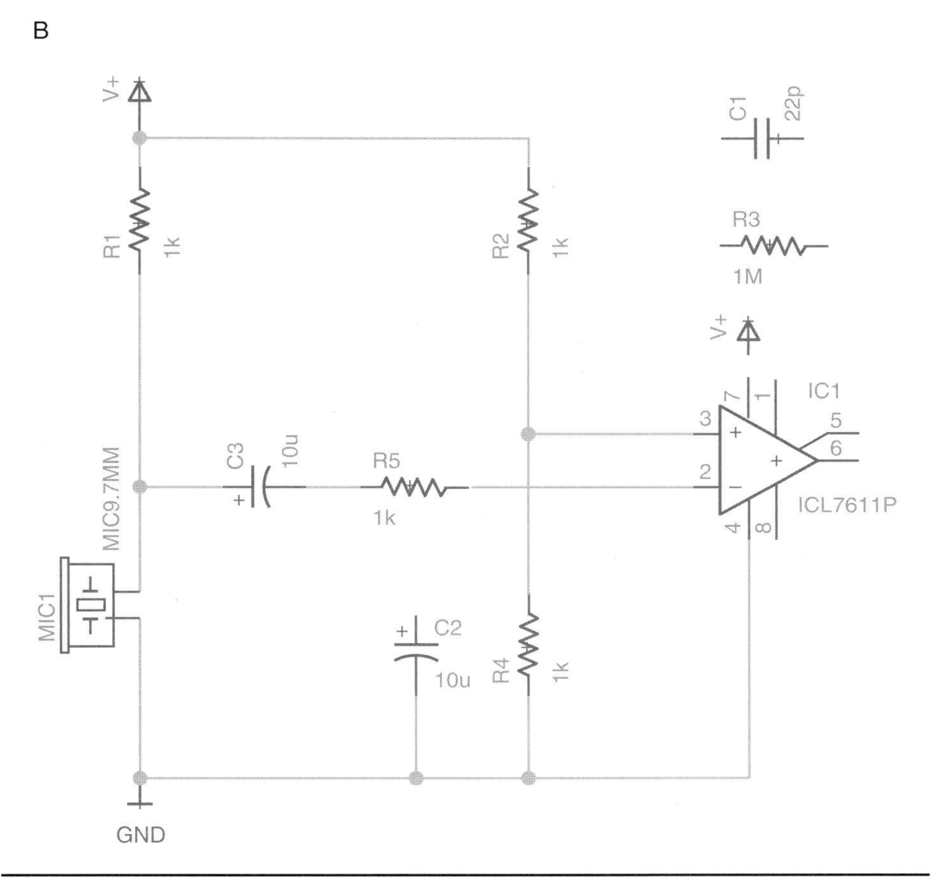

Figure 4-22 Connecting up the amplifier section (*continued*).

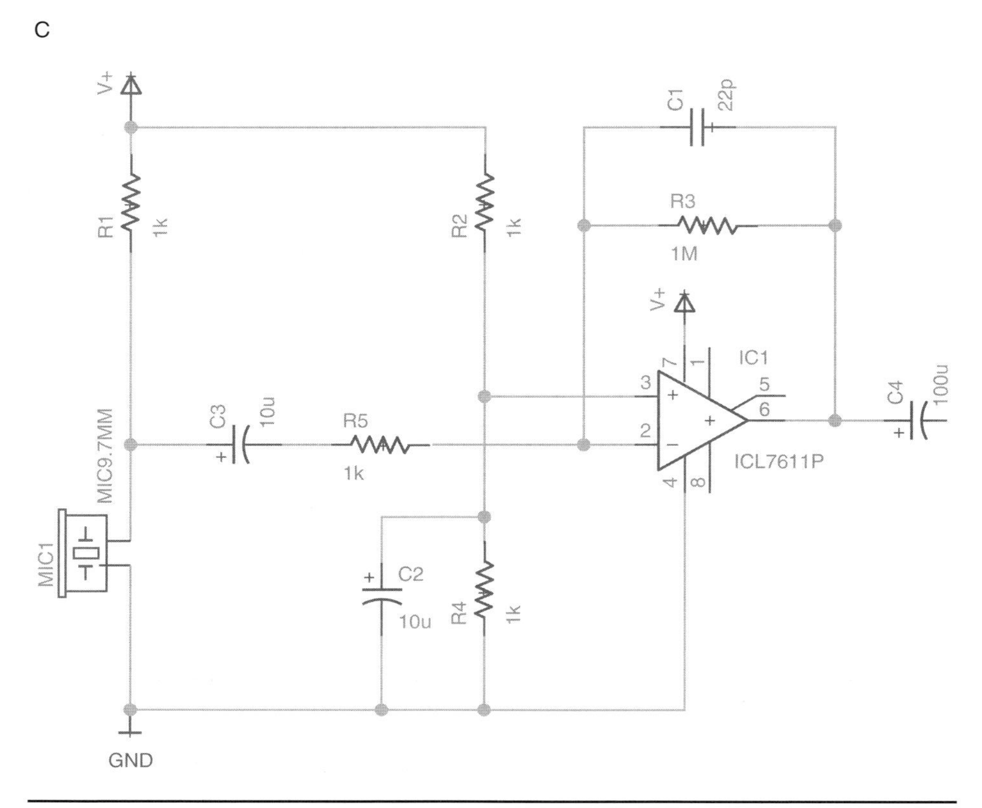

Figure 4-22 Connecting up the amplifier section (*continued*).

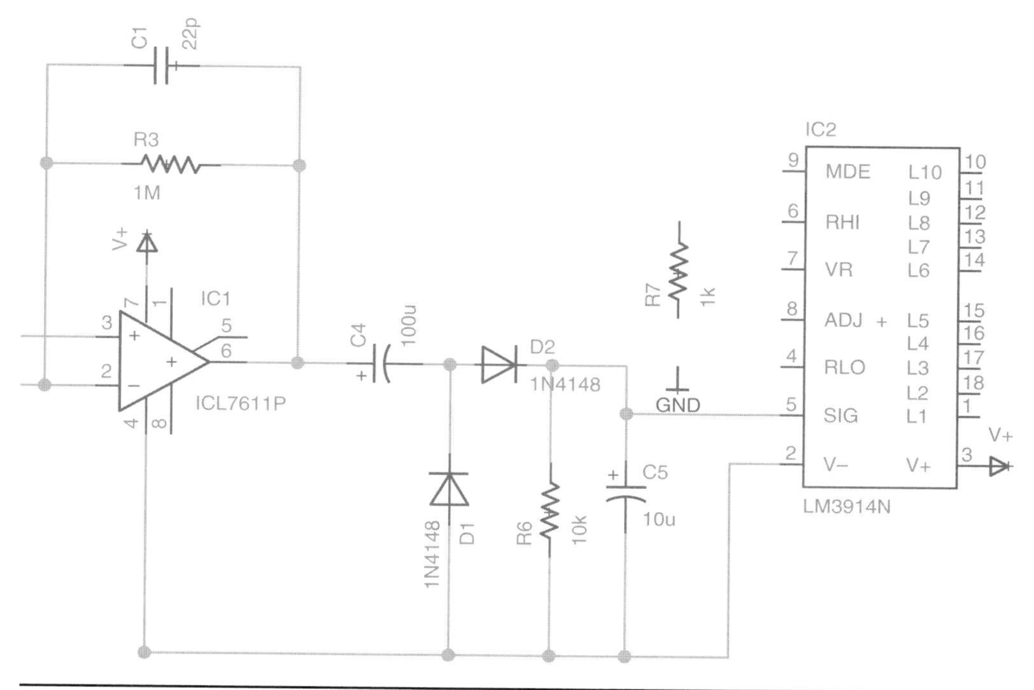

Figure 4-23 Connecting up the filter section.

Next, continuing to the right of the schematic, we wire up the filter section so that it looks like Figure 4-23.

Note that we have continued the GND net from the amplifier actions. I also had to move the GND supply near pin 4 of IC2 because it was in the way.

The final stage is to connect up the bar-code part of the schematic around IC2. Getting all the nets between the output pins of IC2 and the LEDs can be a bit tricky. You could, if you prefer, use a bus for this (see the earlier section in this chapter). The name of the bus needs to be something like "LEDS:CATHODE[1..10]" to allow the 10 connections to the LED cathodes to be made from IC2. The bus and direct-connection versions of the schematic are shown in Figures 4-24 and 4-25. On balance, I think the direct-connection approach is clearer.

Assigning Net Classes

This step is not essential, but it is a good idea. It allows you to specify what kind of net a net is. We only really have two types of nets in this design—power and signal—but you could imagine a design that controlled a 110-V relay and had some nets that were high voltage and high current. While this has no bearing on the schematic, when we come to lay out the tracks on the board, it would be useful

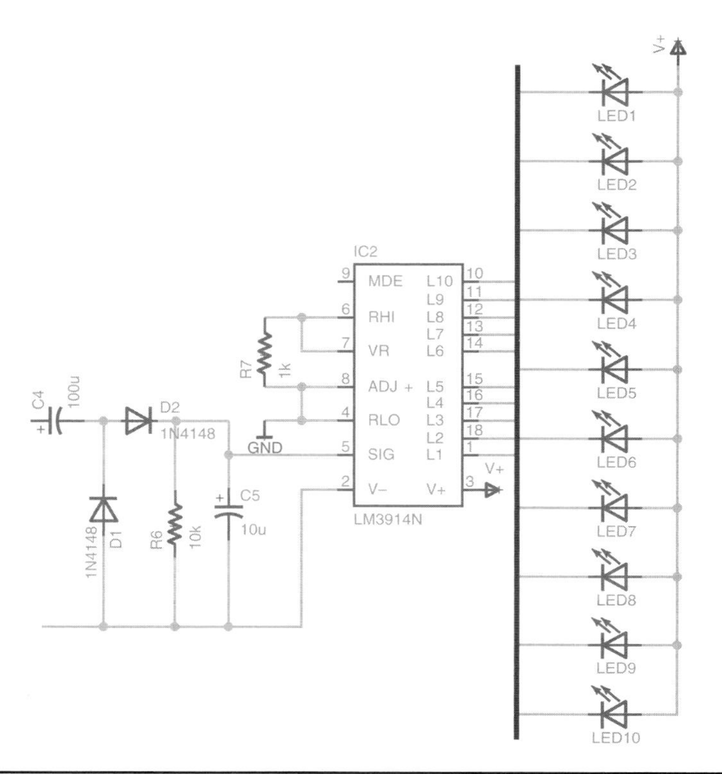

Figure 4-24 Bar-code section of the design—using a bus.

to know that those tracks should be wide and well separated from other tracks. Similarly, it is common to use slightly thicker tracks for supply nets.

As the design currently stands, all the nets will be of a class called *default*. If you use the Information command and select a few nets, you can see the dropdown near the bottom of the window that allows you to select a class for the net. Currently, there is only one entry there ("Default"). To add another net class for supply, select the option "Net Classes.." from the Edit menu, and type a name in the second row (Figure 4-26).

We will return to the other parameters of a net class in Chapter 5 when we come to lay out this board. For now, just add the name.

Now change the class of the GND and V+ nets to "Supply" using the Information command (Figure 4-27).

This completes the connecting up of the nets. The next step is to validate the schematic using the ERC.

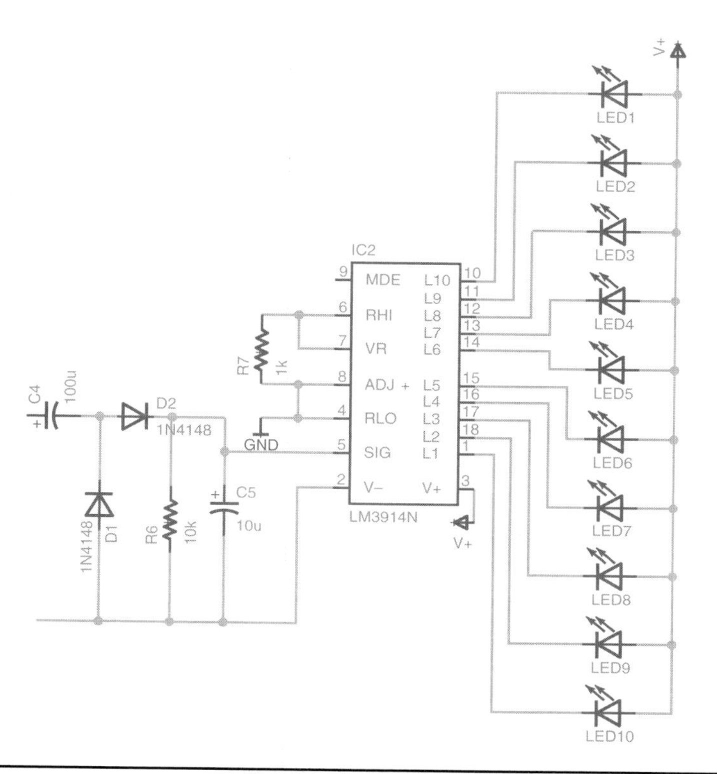

FIGURE 4-25 Bar-code section of the design—direct connections.

Net classes

Nr	Name	Width	Drill	Clearance
0	default	0mil	0mil	0mil
1	supply	0mil	0mil	0mil
2		0mil	0mil	0mil
3		0mil	0mil	0mil
4		0mil	0mil	0mil
5		0mil	0mil	0mil
6		0mil	0mil	0mil
7		0mil	0mil	0mil
8		0mil	0mil	0mil
9		0mil	0mil	0mil
10		0mil	0mil	0mil
11		0mil	0mil	0mil
12		0mil	0mil	0mil
13		0mil	0mil	0mil
14		0mil	0mil	0mil
15		0mil	0mil	0mil

OK >> Cancel

FIGURE 4-26 Defining a new net class.

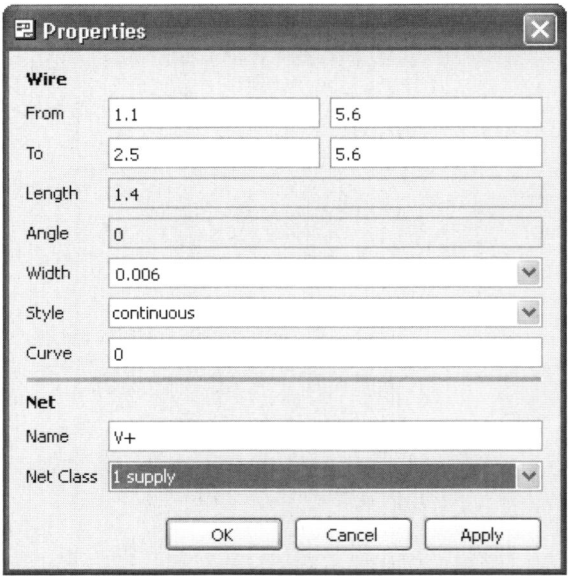

FIGURE 4-27 Using the Information command to change the net class.

Running the ERC

Before we start on the layout, we need to run the ERC. This will tell us about any problems in the schematic. It is important to understand that the ERC does not simulate our circuit; if the design is wrong, it will not tell us that it is wrong. It basically just checks that there are no dangling connections that have been missed or any nets that run too close to each other.

It is surprisingly easy to create a schematic in EAGLE that looks like everything is connected just fine, but it turns out that there are some connections that do not quite meet.

Launch the ERC, either from the Tools menu or from the ERC command, bottom left of the Command bar. The result should look like Figure 4-28. The ERC is very helpful because if you select an item on the list, it will highlight the relevant area on the schematic.

As you can see from Figure 4-28, the ERC is only actually reporting one error. The error is that IC2 has an unconnected pin (pin 9). You may have noticed that pins 1 and 8 of IC1 are not connected to anything either, yet the ERC is not complaining about them. This so because the part for IC1 from the library defines those pins as being allowed to be unconnected, whereas the definition for IC2 says that pin 9 must be connected.

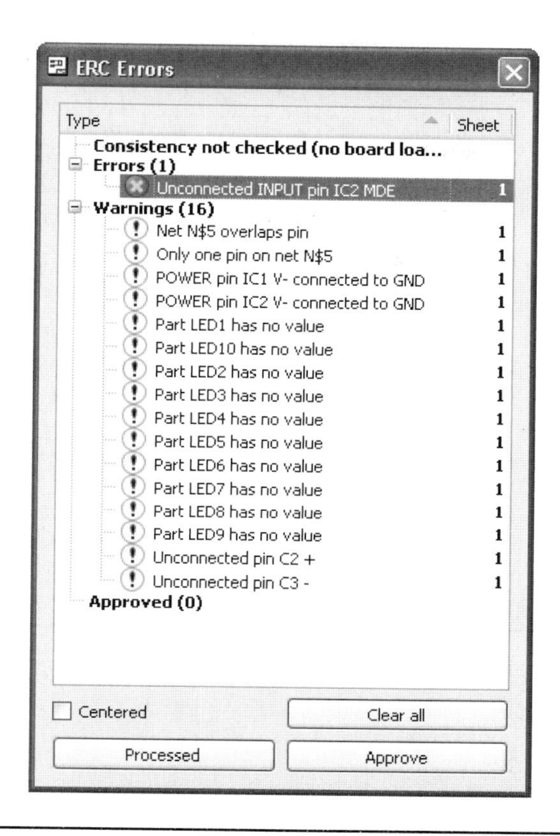

Figure 4-28 ERC results.

Looking at the datasheet for the LM3914, it states that if you want the bar-code driver to operate in "dot" mode, then pin 9 should be left unconnected. Therefore, actually, the definition for the IC2 part is incorrect. In Chapter 11 we will look at how you can copy parts from one library to your own library and modify them. For now, though, we are going to allow this error to stand because we know that it is not a real error but just a bug in the part definition.

In the warnings section, if you have been careful connecting up your nets, you will not have a warning like the first one: `Net N$5 overlaps pin`.

To see what this warning refers to, click on the warning in the list, and a line will spring forth pointing to the part of the schematic where the problem is located (Figure 4-29).

What this error means is that when I was drawing the net between C3 and R5, I started too far to the left on C3 and missed the lead. This means that C3 is not actually connected to R5, which is why there is a second warning that says `Only one pin on net N$5`.

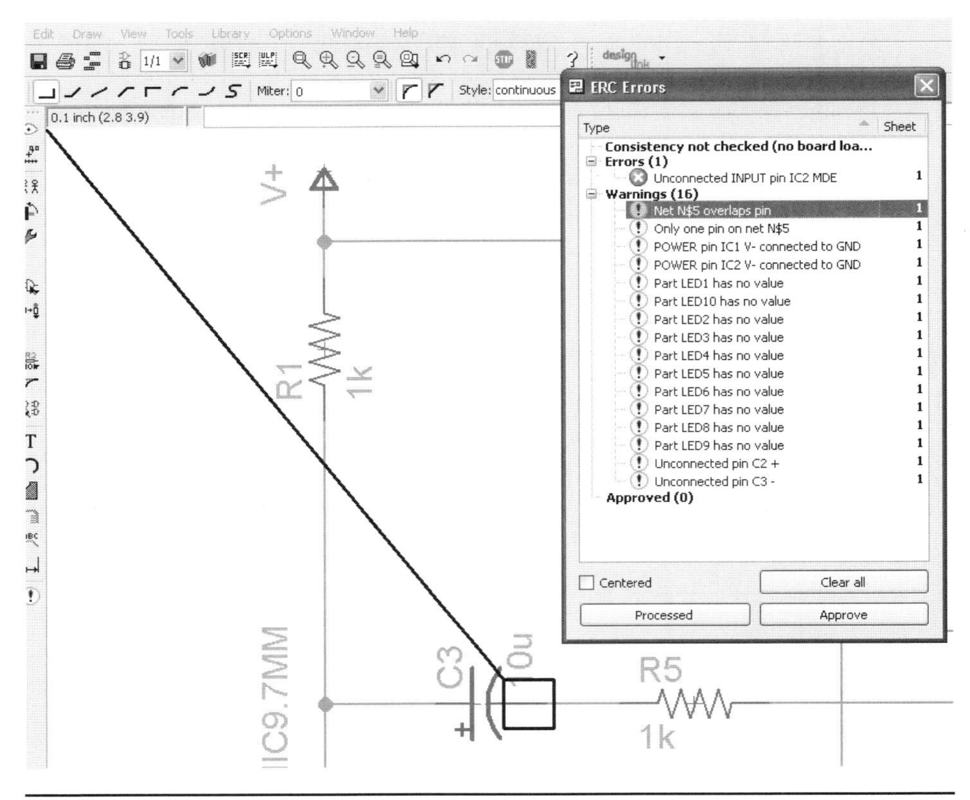

Figure 4-29 Net overlapping pin warning.

See what I mean about a schematic looking correct but not actually being connected up right? This is why it is very important to understand every one of the ERC results and either fix them if there is genuinely something wrong (as in this case) or ignore them if it is safe to do so (as in the case of the unconnected input on IC2). Once you get to laying out the PCB, it may not be obvious that there is a connection missing, and you may end up fabricating boards that are useless.

The best way to fix this is to delete the net and draw it again. Thus, close the ERC results window, select the Delete tool, click on the net in between C3 and R5, select the Net command, and then carefully draw the net in again from the right-hand lead of C3 to the left-hand lead of R5. Run the ERC again to make sure that it is fixed.

The next two warnings are just about the naming of power nets. Both IC1 and IC2 expect the negative supply pins to be called V–, and we are connecting both to GND. This can safely be ignored.

The next 10 warnings (one for each LED) just tell us that we have not given the LEDs values. They don't need values, although you could use the "Value" field to specify the color, say, but we will chose to ignore these warnings.

The final warning in the list shows that we have just plain forgotten to connect the top end of C2 to anything. This is easily remedied.

Summary

EAGLE is a very sophisticated tool. Although we have built on the simplified schematic design process that we first started in Chapter 2, there are still features of EAGLE that we have not used. We have, however, learned how to use all the features that you will need to use for a fairly straightforward schematic diagram.

In Chapter 5, we will give an introduction to how to use the Board Editor in general and then work through the bar-code example laying out both fully through-hole and surface-mount versions of the PCB design.

CHAPTER 5
Laying Out a Printed Circuit Board

In this chapter, you will first look at the various commands available to use when laying out a PCB and then make two example layouts for the bar-code example of Chapter 4. One of these layouts will use a typical hobbyist's through-hole design. The other will use mostly surface-mount components. In the final section of this chapter, we spurn the help of the autorouter, lay out the same example manually, and compare it with the automatically generated layout.

Experimenting

Before starting on the sound meter example, you will be experimenting with the many features that are available through the Board Editor. To have something to experiment with, use the board from the flasher example of Chapter 2 (Figure 5-1). During your experimentation, you are quite likely to completely mess up these project files, so remember that the original files can always be downloaded again from the book's website (www.simonmonk.org).

Open the flasher example project, and then open the board layout.

Layers

A fundamental part of the Board Editor is the concept of layers. EAGLE defines a huge set of layers and also allows you to define your own layers and even automatically generate them from scripts. You will only ever need to use quite a small subset of these layers. At any time, you can control which layers are visible and which are hidden. This is useful to avoid clutter and to be able to concentrate

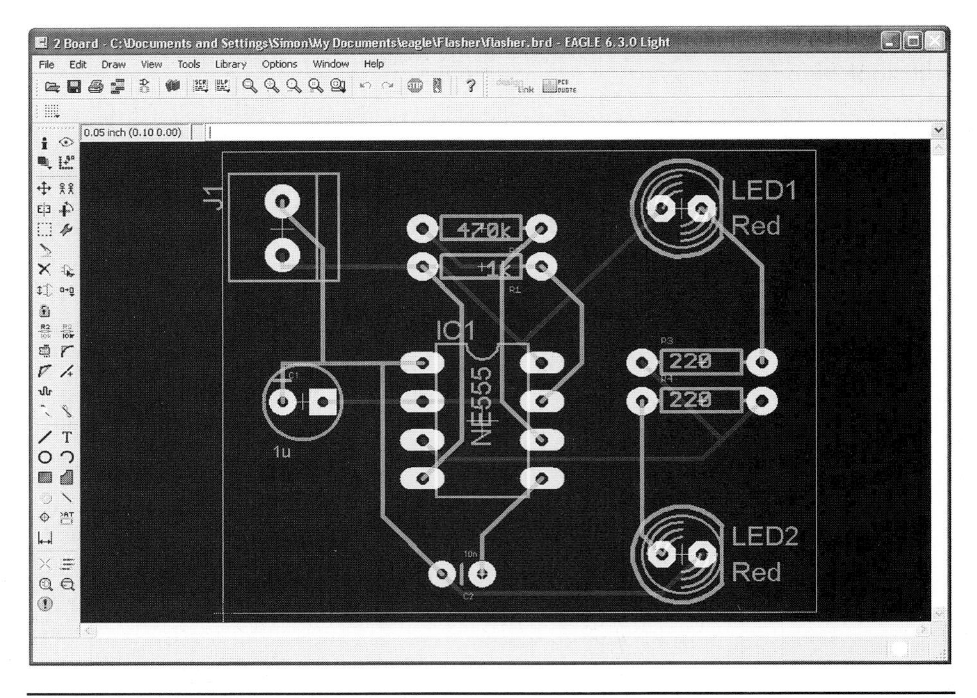

FIGURE 5-1 Board layout from the flasher project.

on a particular aspect of the design without the confusion of everything being visible all the time.

The Layers command (second from the top icon on the left of the Command bar) provides you with a list of all the layers available, with the ones currently visible highlighted (Figure 5-2).

The scroll bar tells you just how many layers there are off the bottom of the window. Let's look at the layers that we will most definitely use before looking at some of the more exotic ones. Table 5-1 lists the most common layers.

When you add a part to a schematic, that part can contribute various things to different layers in EAGLE. Thus part of the package description will be such things as the size and shape of the pads and the names and values that will end up on the "Pads" and "tNames" (or "bNames") layers.

The Command Toolbar

The basic structure of the Board Editor is the same as that of the Schematic Editor. The Command window on the left has commands, some of which, such as Move

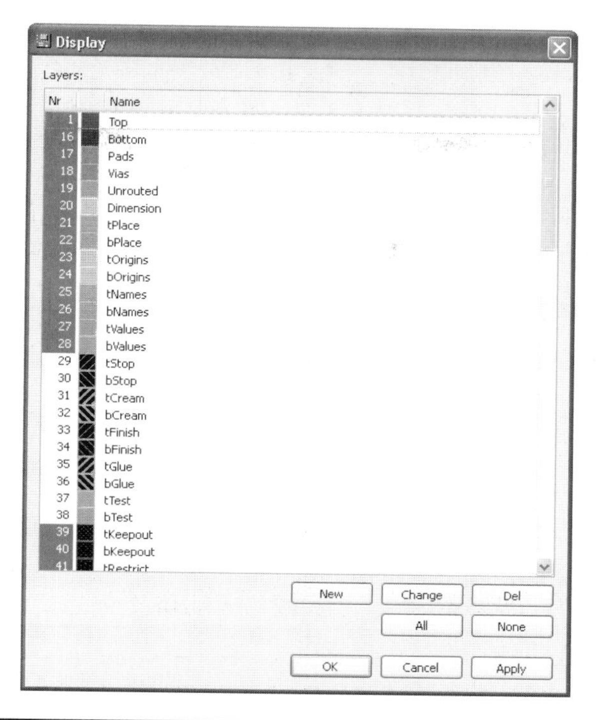

FIGURE 5-2 Layers.

TABLE 5-1 Main Layers in Board Editing

Name	Default Color	Description
Top	Red	The top layer of copper track
Bottom	Blue	The bottom layer of copper track
Pads	Green	Copper pads to which components are soldered
Vias	Green	Links from top copper layer to bottom copper layer
Unrouted	Muddy yellow	Net links (air wires) that have yet to be turned into copper tracks
Dimension	Light gray	The size of the PCB
tPlace	Gray	The outline of the components on the top of the board
bPlace	Gray	The outline of components (if any) on the bottom of the board
tNames	Gray	The names of components on the top of the board (R1, C1, etc.)
bNames	Gray	The names of components (if any) on the bottom of the board (R1, C1, etc.)
tValues	Gray	The values of components on the top of the board (1 kΩ, 10 nF, etc.)
bValues	Gray	The values of components (if any) on the bottom of the board (1 kΩ, 10 nF, etc.)

and Information, are the same as when using the Schematic Editor and some of which are new.

Figure 5-3 shows the Command toolbar for the Board Editor. The most important commands are highlighted in bold, and the ones marked with an asterisk (*) are new commands that only apply to the Board Editor and not the Schematic Editor.

Common Commands

Many of the commands in the Board Editor work in just the same way as those in the Schematic Editor. In this section we will look at just the most common commands.

Move

The Move command allows you to move parts around on the board layout. It does not affect the position of the same part in the Schematic Editor.

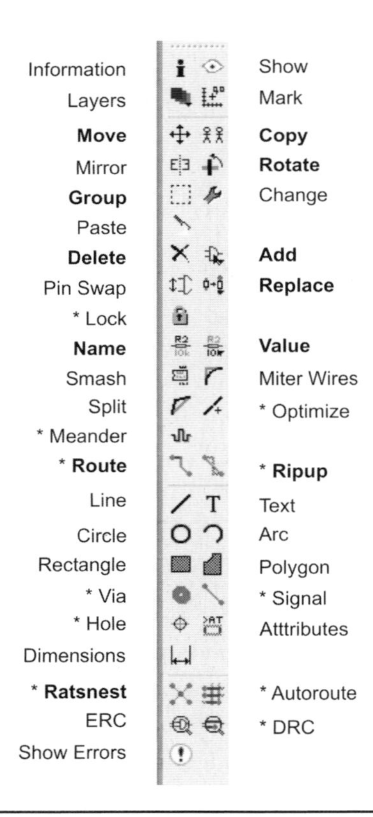

FIGURE 5-3 Board Editor Command toolbar.

Group

Group moves and deletes are carried out in the same way as in the Schematic Editor. That is, you click the command (say, Move) and then click the Group command, drag over an area to make a multiple selection, and then right-click and use the Group move (or whatever) option in the menu. When laying out by hand, a very common Group operation is to Ripup a section of tracks that have been routed so that they can be rerouted better.

Delete

I have indicated that this is an important command, but actually it's a command that is important that you do *not* use in the Board Editor, at least on parts. If you want to delete parts, then you should do it in the Schematic Editor. When you delete them there, they will automatically disappear from the Board Editor. If you try to delete a part in the Board Editor, you will see the error message in Figure 5-4.

The message is telling you that the change you are trying to make to the board cannot be applied back to the schematic. The exception to this is that simple changes such as changing the value of a component can be made in the Board Editor, and the change will then also appear in the Schematic Editor. Generally, it is better to assume that the schematic is the important master copy and that any design changes should be made there.

Name

You can safely change the name of a part in the Board Editor. Such a change will automatically update the corresponding part in the Schematic Editor.

Route

The Route command is what you use to convert an air wire (connection needing to be made) into a copper track. We will use the Route command extensively later on when we start laying out our example board.

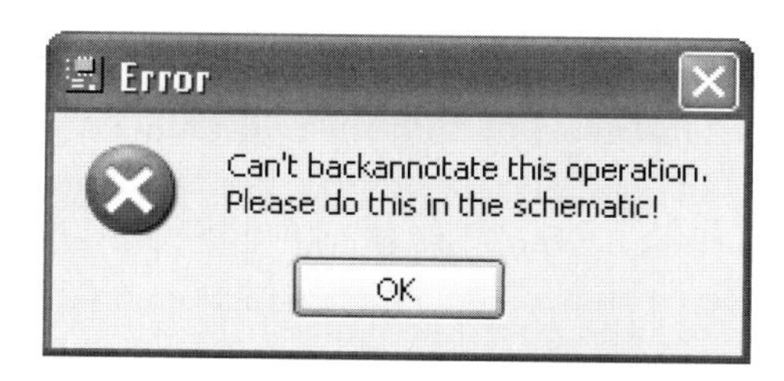

FIGURE 5-4 Trying to delete a part in the Board Editor.

Figure 5-5 Routing Parameter toolbar.

When the Route command is selected, the Parameter toolbar fills with all sorts of useful parameters that we can set for the copper track we are about to route. We will see later that by using design rules and net classes, these may get set for us automatically, although there is always the option to tweak them manually. Figure 5-5 shows the Routing Parameter toolbar.

From left to right, the icons are

- **Grid.** More on this later.
- **Layer.** Copper on the top or bottom of the board.
- The next eight icons represent different styles of track bend. To get the hang of these, you will experiment with them when we try laying out the bar-code example.
- There is then a "Miter" dropdown that allows you to select a curve radius for bends in the track. This used to be important for certain chemical etching processes where tight bends could cause the etchant to collect and overetch the track. Unless you are home etching your PCBs, this feature is largely down to personal preference.
- Finally, there are two selection buttons that allow you to choose between square and curved miters.

Ratsnest

If you are routing a PCB by hand, you will probably find yourself hitting this button frequently. When clicked, it recalculates the shortest routes for all the air wires. Remember that these air wires indicate where there is more routing to do. When all the connections have been made, it will display the status message `Nothing to do!`

If there are still connections to be made, this command will recalculate them, indicating the shortest routes between existing tracks. Thus, having done a bit of routing, clicking Ratsnest will tidy everything up and recalculate the air wires for you.

Copy

If you remember from the Schematic Editor, Copy is actually more like "duplicate." It is another command that should not be used with parts. Switch to the Schematic Editor if you want to add or duplicate parts. It is, however, useful to copy board-specific items that do not appear on the schematic, such as holes or text.

Rotate

This rotates the part through the angle specified in the Parameters toolbar. This is normally 90 degrees. However, you can use other angles by

- Clicking the Rotate command
- Typing the angle into the dropdown (say, 45) and hitting "Return"
- Clicking Rotate again
- Selecting 45 in the dropdown
- Clicking on the part to rotate

This can be useful for unusually sized PCBs.

Add and Replace

Adding a part is another operation that is not permitted from the Board Editor, and while replacing parts, if they have compatible pins, it is much better to go back to the Schematic Editor and make the change there.

Value

Changing a component value in the Board Editor is just fine. Any change that you make in the Board Editor will automatically update the Schematic Editor.

Ripup

Ripup is the opposite of Route. It allows you to rip up sections of track so that they can be rerouted. You will use this command a lot.

Other Commands

You can, if you wish, skip over the commands in this section. They are described here for completeness and so that you know of their existence should you need them at some point in the future.

Mirror

The Mirror command is only useful if you are placing components on the bottom layer, as well as the top. Mirroring the component to go on the underside will automatically mirror the pads.

Paste

Paste is the same concept as the Paste in Cut and Paste, but this being EAGLE, it works in a different way. To make a copy to be pasted, you select items either individually or using the Group command. The Paste command then will attach a duplicate of everything selected to the cursor so that you can click to place it in the design. As with duplication, you should only use this feature with items that are specific to the board design, such as holes or text.

Pin Swap

Pin Swap is quite handy for making the routing process more convenient. For two-pin devices, you can achieve the same effect just by rotating it, without the side effect of it messing up your schematic (as has happened in Figure 5-6 following a Pin Swap).

Lock

The Lock command allows you to protect a part so that it cannot be moved or modified without first unlocking it. To unlock it, hold down the "Shift" key while using the Lock command. Or you could just be careful and use Undo if you make a mistake.

Smash

This is the same Smash command that lets you separate the labels for a part from the part itself in the Schematic Editor. This can be useful if you want to move, say, a part value away from the outline of the part to prevent overlaps of text and the part.

Split

Split allows you to add an extra bend on the segment of a track without having to reroute the section.

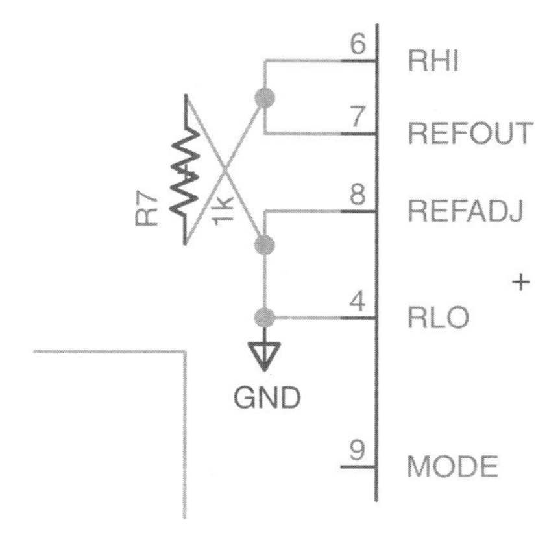

Figure 5-6 Pin-Swapping in the Board Editor.

Meander

Meanders are used in high-frequency projects where the track length can affect when signals arrive at a component. A Meander zigzags to lengthen the track. This is something of a professional feature needed for boards operating at very high frequencies and unlikely to be needed by us hobbyists.

Shapes and Text

There are various different shapes and text that you can use to draw on your board. This can be in copper if you select top or bottom as the layer or can appear in the final silk screen if you select the layer that will eventually be used for silk screen—often tPlace and bPlace.

Remember that parts will come with their own text and placement guides, so you do not usually need to draw much in the way of annotation. The Polygon shape, when drawn in copper, can be used to create a ground plane, something we will discuss in the example section.

Via

Vias are used to allow a track from one layer to continue on another layer, that is, from top to bottom or vice versa. Physically, they are small holes with a connection between the layers. It is better to create vias automatically as you route simply by switching layers than to add them explicitly. In other words, avoid the Via command.

An exception to this is if you have two ground planes (see later in this chapter), one on the top layer and one on the bottom, and you want a good connection between them. You might then add a number of vias to achieve this.

Hole

It is quite common to add a hole to a PCB so that it can be fixed into an enclosure. The Hole command lets you do this. When adding a hole, be careful that no tracks will be cut by the hole.

Dimensions

The Dimension command lets you annotate part of a design with dimensions.

ERC

You should not need this from the Board Editor because you should not be changing things there that affect the electric rule checker (ERC).

Show

This command has a similar purpose to the Information command except that rather than opening a little window that gives you details of the component selected, it displays the information in the status area at the bottom of the screen.

Mark

The Mark command allows you to set a local origin anywhere on the Board Editor so that you can see the coordinate values relative to that point. Clicking on the "GO" button (looks like a traffic signal) cancels the Mark. This can be used to measure things on the board, but frankly, I have never found a good use for it.

Change

The Change command allows you to change almost anything about an object on the schematic diagram. You can, of course, change these things using the Information command.

Miter Wires

This command allows you to put a curve in a line or polygon. To use it, select the command, and then click on a corner of a line. You can then set the radius in the dropdown in the Parameter toolbar.

Optimize

This removes unnecessary midpoints from a straight line.

Signal

The Signal command would allow you to define signals in the board layout, but if you try to click on it, EAGLE just tells you that you can't use it. It will only work if forward/backward annotation is disabled, which would be silly. Just ignore this feature.

Attributes

You can add your own custom attributes to a component and then decide whether you want them to be displayed. See Chapter 4 for an example of using this to add a "power" property to a resistor.

Autoroute

This command runs the autorouter. More on this later.

DRC

The design rule checker (DRC) is to the Board Editor what the ERC is to the Schematic Editor. It performs all sorts of checks that make sure that tracks are not crossing each other on the same layer, located too close together, and so on.

Because the design rules are complicated to set up, you can load and save sets of design rules and even download sets of design rules that others, such as Sparkfun, have made available to everyone.

The Grid

The schematic is drawn on a grid, which helps you to lay things out neatly by allowing you to snap parts and net lines to the nearest point on the grid. However, the grid is much more important for board design.

By default, the grid spacing is 0.05 in. You can change this using the Grid option on the View menu. This opens the dialog shown in Figure 5-7.

Occasionally, it can be useful to reduce the grid spacing if you are running short on space. It is a good idea to keep it at fractions of 0.1 in. because most parts have leads on that pitch.

Sound Meter Layout (Through-Hole)

Now that we have taken the tour, it's time to start working on a board. We are going to begin with a through-hole board design for the sound meter example project.

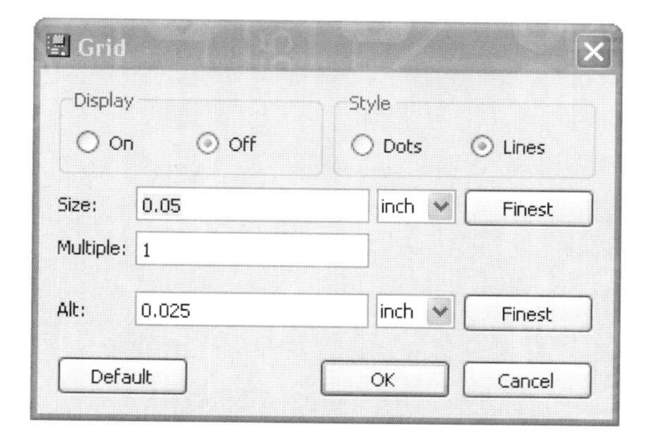

FIGURE 5-7 Changing the grid.

Rather than show you a perfect first-time layout, this design will lead you through the real example as I worked through it. This is not a perfect example layout. There are deliberate initial errors, which we will work through and correct, that are quite likely to crop up when you do your own first designs.

Open the schematic diagram for the sound meter project.

Create a Board from the Schematic

We do not yet have a board for this schematic, so from the File menu, select the option "Switch to Board." You will be informed that the board does not exist and will be asked if you want to create it from the schematic. Say "Yes," and the Board Editor will open, looking something like Figure 5-8.

Thus we have what could be described as a big heap of components on the left and a rectangle representing the board on the right. We even have some text that has escaped from the frame on the schematic. Delete the block of text to make it easier to concentrate on the components.

Decide on Board Size

Sometimes you will know what size board you are aiming for, and other times it is best to add the components and do some of the layout before making a decision. Because 5 by 5 cm (just short of 2 by 2 in.) is used as the cutoff size for a big price hike by the author's favorite PCB fabrication shop, we will give ourselves the constraint of fitting the design onto a 5- by 5-cm board.

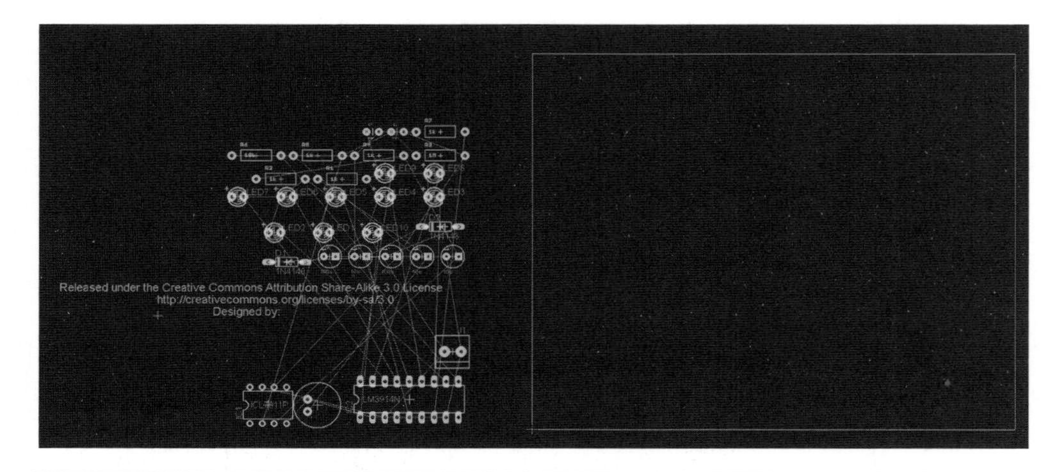

Figure 5-8 An initial board.

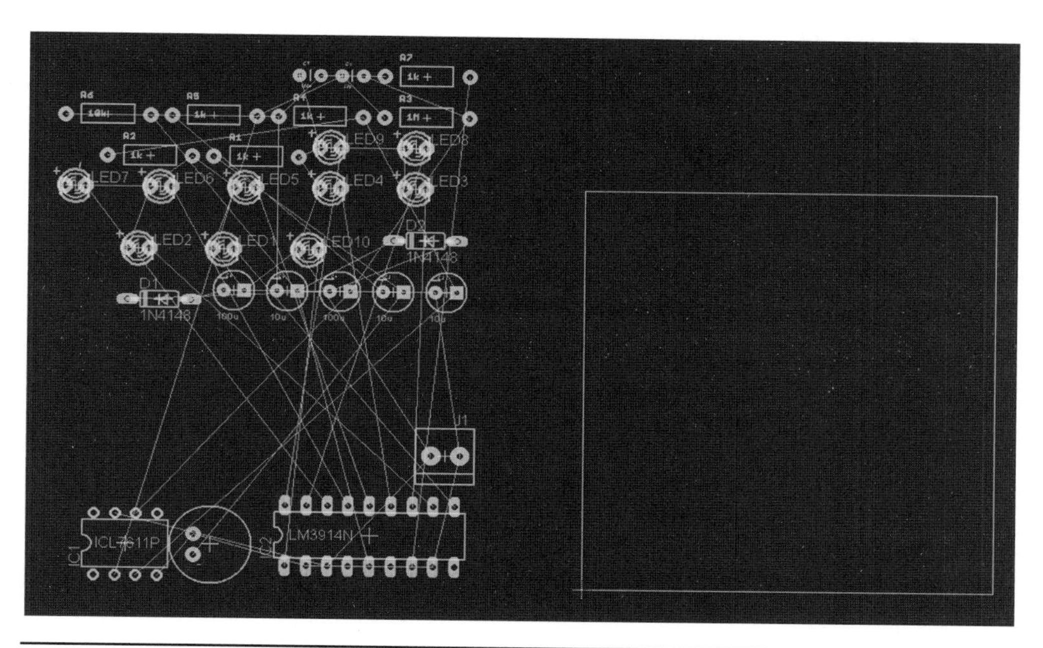

Figure 5-9 A quart and a pint pot.

Our units are all in inches, so move the top and right edges (drag them in the middle) so that the board is 1.95 in. square. Watch the size change in the top left of the editor area.

At first sight, it does not look like everything is going to fit on the resized board (Figure 5-9), but trust me, it will.

Position and Rotate the Components

Let's start moving the parts onto the board. The layout broadly follows the schematic. Thus we can put IC1 and associated components over on the left. Then the filter components and IC2 and, finally, the LEDs can be put over on the right side.

Remember that when using the Move command on a part, you also can rotate it through 90 degrees by right-clicking. Frequent use of the Ratsnest command will tidy up the air wires. You will probably also need to zoom in to be able to identify the parts correctly.

In deciding where to place the components so as to broadly follow the schematic, it is very helpful to have the schematic available for reference. If you have two monitors, keep the schematic up on one of them and the Board Editor on the other or simply print out a copy of the schematic to refer to as you place the components.

I used the Smash command on all the capacitors and then used the Change (size 0.05) on the part name label to make it easier to read the capacitor names. We will come back to part labels later in this chapter when we sort out the silk-screen text.

When all the components are in place, the board should look like Figure 5-10.

Currently, there is a rat's nest of unrouted wires. Clicking on the Ratsnest command will tidy this up a little (Figure 5-11).

Add Mounting Holes

It will be useful if this board has some mounting holes on it so that it can be attached to a front panel or otherwise secured in an enclosure. Now is a good time to add the holes, before we start routing the air wires, because the tracks will have to stay well clear of the holes.

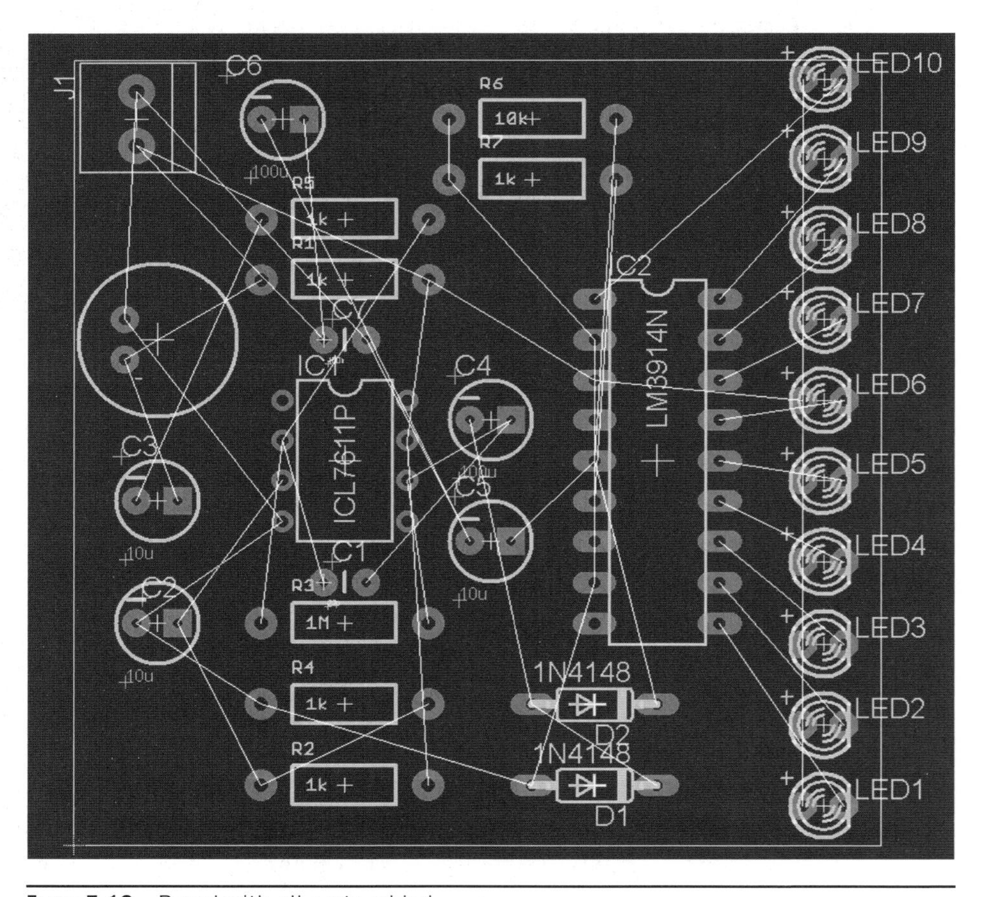

FIGURE 5-10 Board with all parts added.

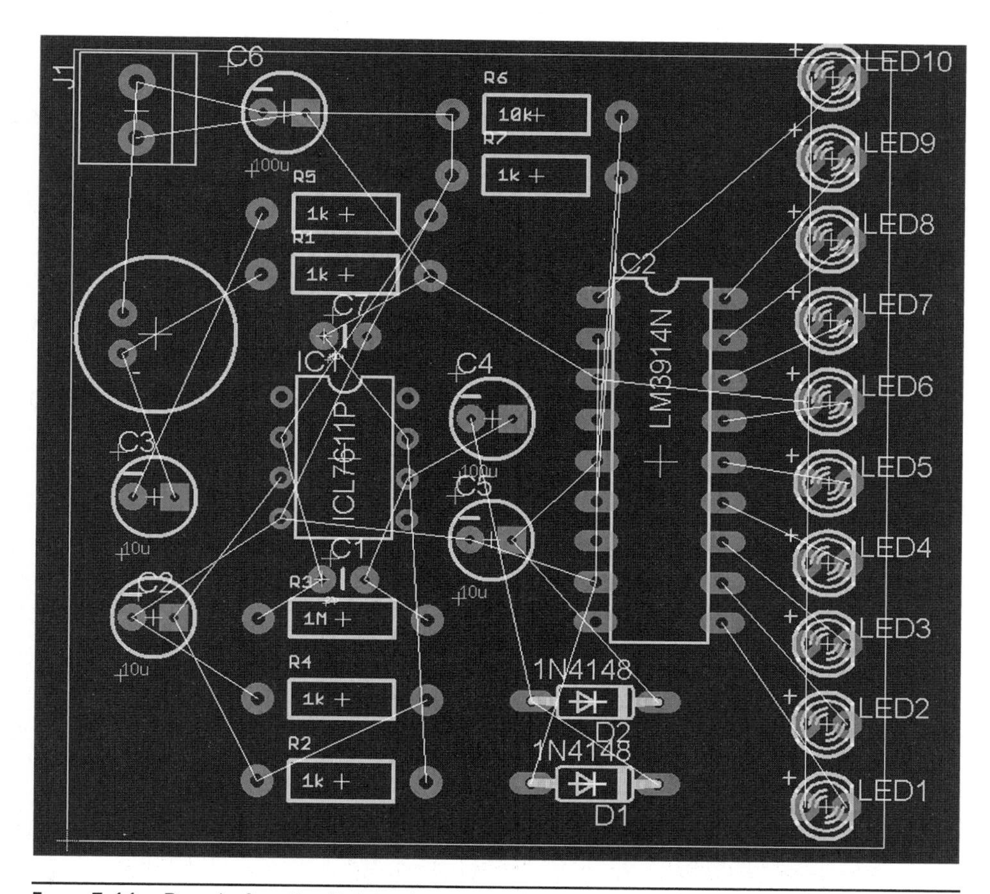

FIGURE 5-11 Board after the Ratsnest command.

Select the Hole command from the Command toolbar. Change the drill size to 0.2 in. in the dropdown list in the Property toolbar, and then add three holes, as shown in Figure 5-12. It does not matter if the holes are on top of air wires.

Get Some Design Rules

Before we run the autorouter, we need to set up a few rules about how thick we want the tracks to be, how far apart they are to be, and so on. We can do this in two ways, one by using a set of design rules and the other by defining some net classes (which we started in Chapter 4).

Let's start with the design rules. We are going to use a set of design rules that Sparkfun supplies as a free download. You should have installed these back in Chapter 1 (see the section "Installing the Sparkfun Design Rules"). If you did not install them, please do so now.

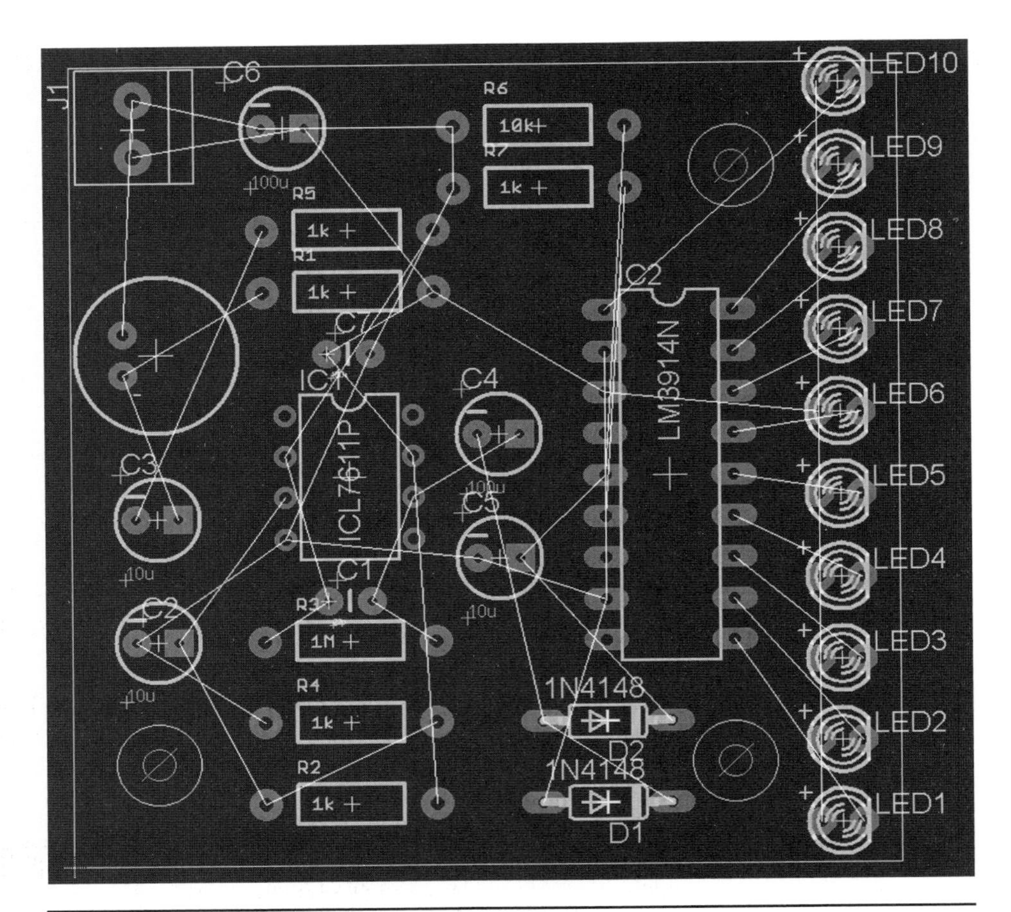

Figure 5-12 Board with mounting holes added.

You may remember that back in Chapter 4 we defined a new net class called Supply, with the intention that it should provide thicker tracks for tracks that supply power to the chips. We created the names for the classes and changed the GND and V+ nets to be of the Supply class, but we did not set the track thicknesses.

Thus, from the Edit menu, select "Net Classes.." and modify the parameters so that they look like Figure 5-13.

This sets the track width to 10 mils (0.01 in.) for a regular signal and 20 mils for a supply signal.

Run the Autorouter

It is time at last for the really fun bit—running the autorouter. Click on the Autoroute command. We do not need to change anything on the Autoroute

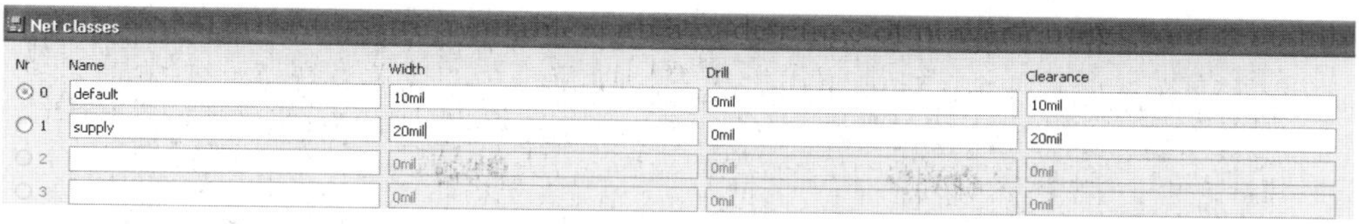

Net classes

Nr	Name	Width	Drill	Clearance
0	default	10mil	0mil	10mil
1	supply	20mil	0mil	20mil
2		0mil	0mil	0mil
3		0mil	0mil	0mil

FIGURE 5-13 Setting dimensions in the net classes.

settings; the defaults should work just fine. But let's take a moment to look at some of the settings (Figure 5-14).

This is actually the first of many autorouter settings tabs and probably the only tab that you might want to change. The Preferred Directions setting acts as a hint as to how to use the two layers. By setting one to be predominantly horizontal and the other vertical, you tend to minimize the number of vias you need. If you want to produce a single-layer board, then you can do that here by setting the top layer to be "N/A."

The routing grid specifies the grid to which the bend points in the tracks should be matched. Reducing this number sometimes may be necessary to fit all the tracks in.

Running the autorouter will produce a result something like Figure 5-15. It will not be exactly like Figure 5-15 because your parts will not be in exactly the same positions as mine.

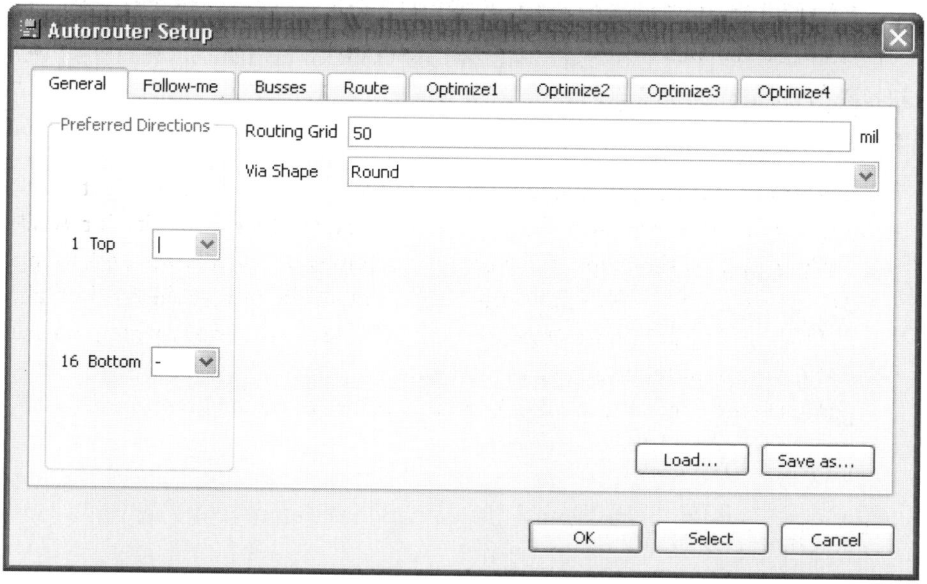

FIGURE 5-14 Autorouter settings.

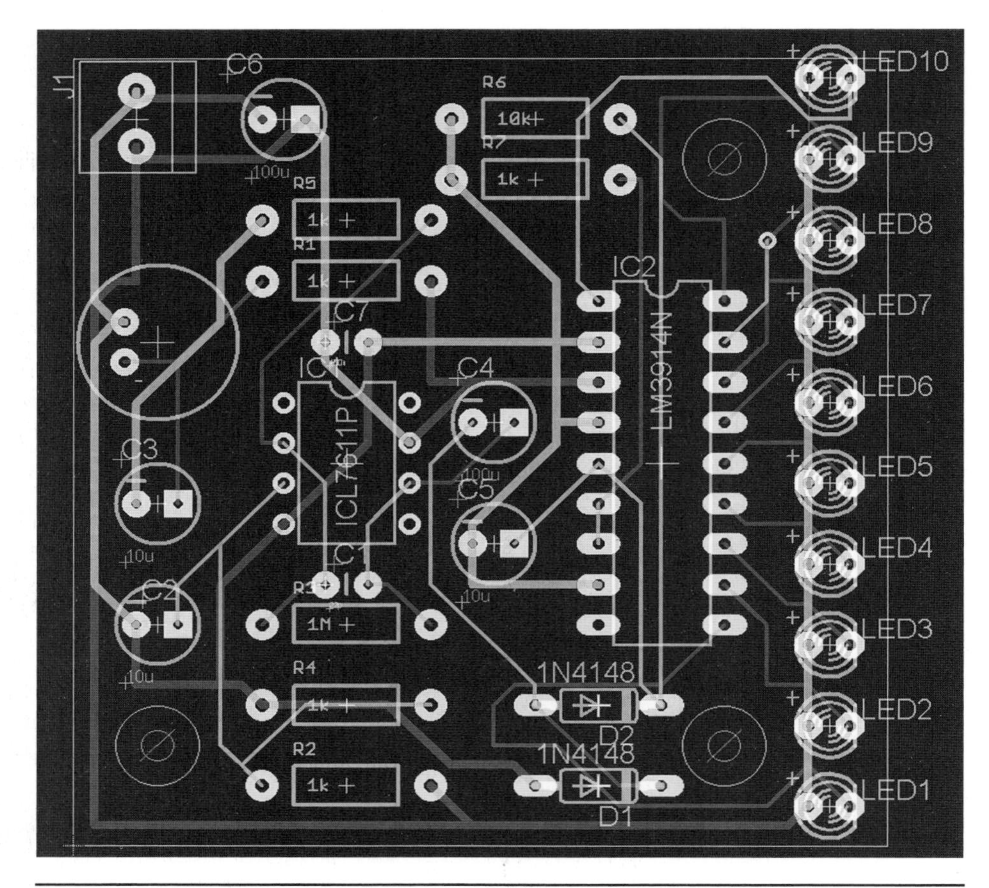

FIGURE 5-15 The first routing.

This is actually a pretty good result. The board looks pretty neat. You can see the thicker tracks for GND and V+. However, you will notice that there are still some air wires that have not been routed.

Tweak the Result

Rather than route the air wires manually, to improve the layout marginally, we are going to add a ground plane. A ground plane, as the name suggests, adds a large area of copper on the board that is connected to ground. Because quite a lot of the components have a connection to ground, this actually reduces the number of tracks required on the bottom layer, making it more likely that all the air wires will route.

Before we can add the ground plane, we need to "rip up" the tracks that have just been laid. You can do this either by using Undo or you can select the Ripup command, select the Group command, drag over the whole board, and then from the right-click menu select "Group Ripup."

Having turned all the tracks back into air wires, add a ground plane by selecting the Polygon tool and drawing a square around the outline of the board (Figure 5-16). Make sure that the Layer dropdown is set to "16 Bottom" and the Width dropdown is set to 0.01 (both on the Parameter toolbar once "Polygon" is selected).

Although the polygon is a shape, it is deemed to have edges, and the "Width" attribute sets the width of these. These are judged by the same design rules as other tracks, and making them wide interacts with the design rules, making the ground plane smaller than the board outline. In short, keep the width small to avoid problems.

When the square has been drawn, there will be a thin blue dashed line all around the outline of the board. Click on "Ratsnest," and the polygon will fill with blue (Figure 5-16).

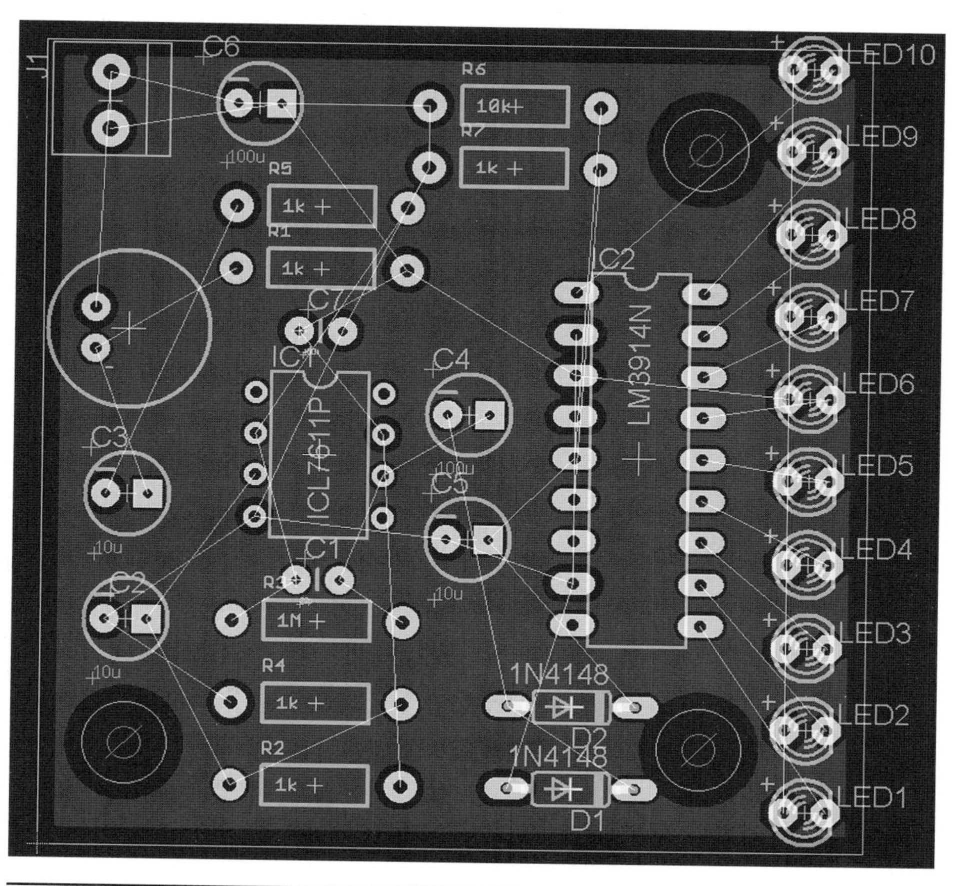

Figure 5-16 Adding a ground plane.

FIGURE 5-17 Naming the ground plane.

Now we need to associate the polygon with the GND signal. To do this, click the Name tool, select "Polygon," and name it GND (Figure 5-17). Note that it can be quite tricky to select the polygon rather than the board dimension lines. When you click on "Ratsnest" again, you will see some of the air wires vanish because they are now taken care of by the ground plane and will not require Track to route them.

Having made the ground plane, run the autorouter again. Figure 5-18 shows the result.

This time the autorouter failed to route just one air wire. Also, the ground plane has broken up. There is a portion over on the right where the ground plane could not spread though because of the design rules. Let's try to sort out these problems.

If you cannot see where the air wire is, then it helps to temporarily hide some of the layers. The air wires are all on the "19 Unrouted" layer.

The remaining air wire is between the two anodes of LED9 and LED10. All the positive leads of the LEDs are connected together on the left. It would be easier if they were on the right because they would then be on the same side as the connections from IC2. So let's rip up all the tracks, rotate the LEDs, and try again. Laying out a board is an iterative process. Unless you have a few components on a large board, you are likely to have to make several attempts to route the board.

This time, the result is much better (Figure 5-19). There is still one air wire between LED9 and LED10, but the ground plane now covers the whole bottom layer.

This looks like it should have been autorouted, and this is probably so because LED10 is so close to the edge of the board that routing it would break a design rule.

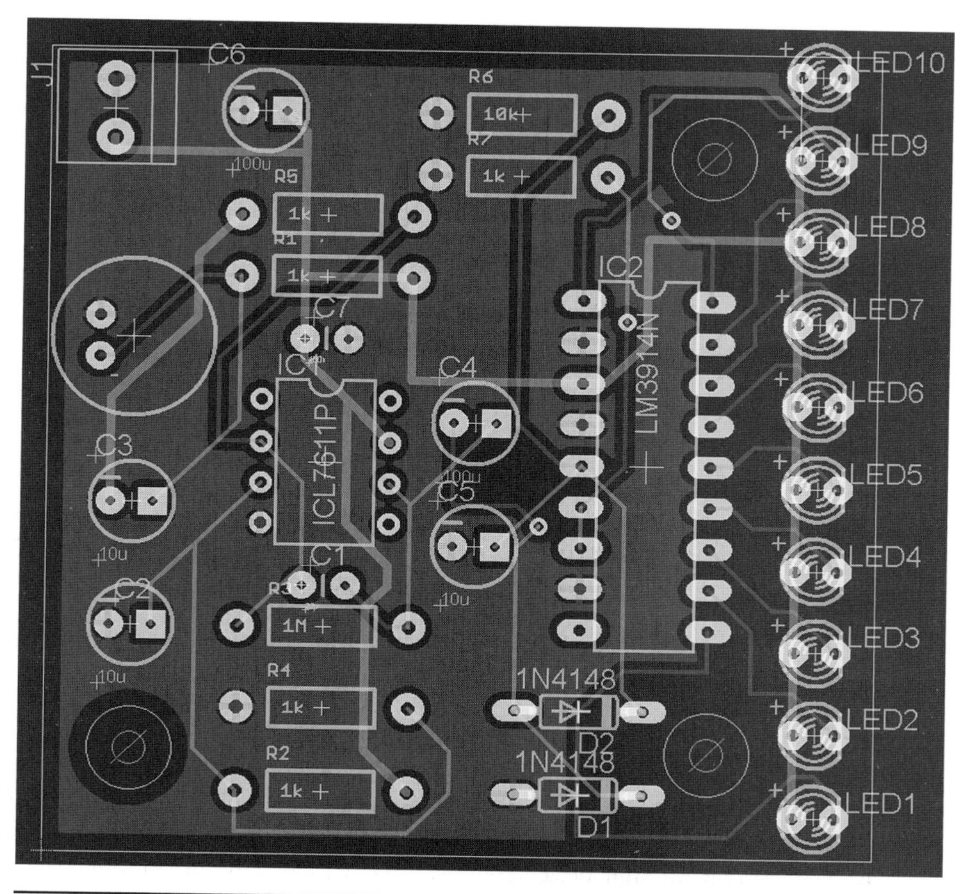

FIGURE 5-18 Autorouting with a ground plane.

Anyway, for now, we will route it manually, just to illustrate manual routing. Select the Routing command, then change the width to 0.2, and make sure that the top layer is selected. Then draw the track up from the positive (+) connection on LED9 to LED10. Figure 5-20 shows this in progress.

Click "Ratsnest" again, and you should be told `Nothing to do`. The full board is shown in Figure 5-21. At this stage, you can probably tell that we will need to revisit the layout soon.

Run the Design Rule Checker

Before we can declare our board finished, we need to run the DRC on it. When this is run on the design, it produces a results window in which you can select problems, and they are highlighted in the design.

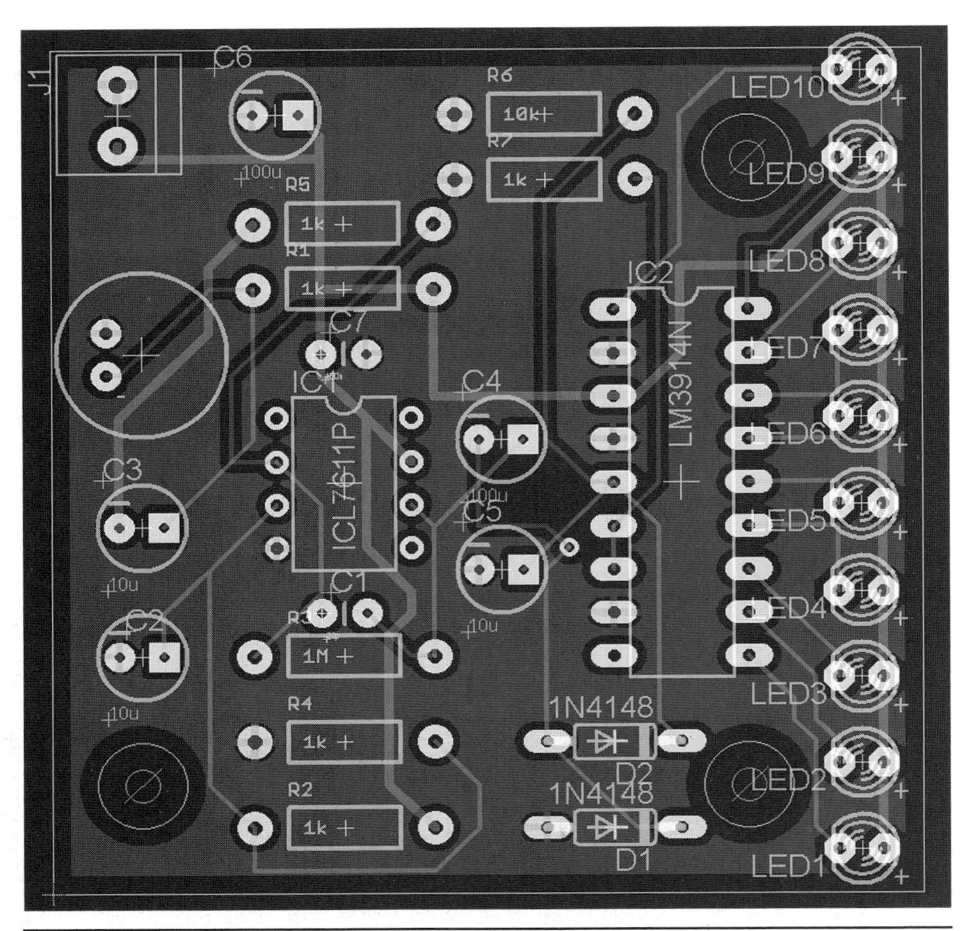

Figure 5-19 A second autorouting with ground plane.

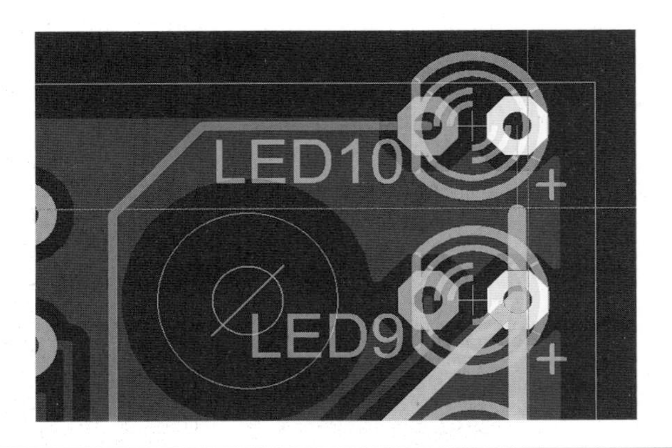

Figure 5-20 Adding the final track.

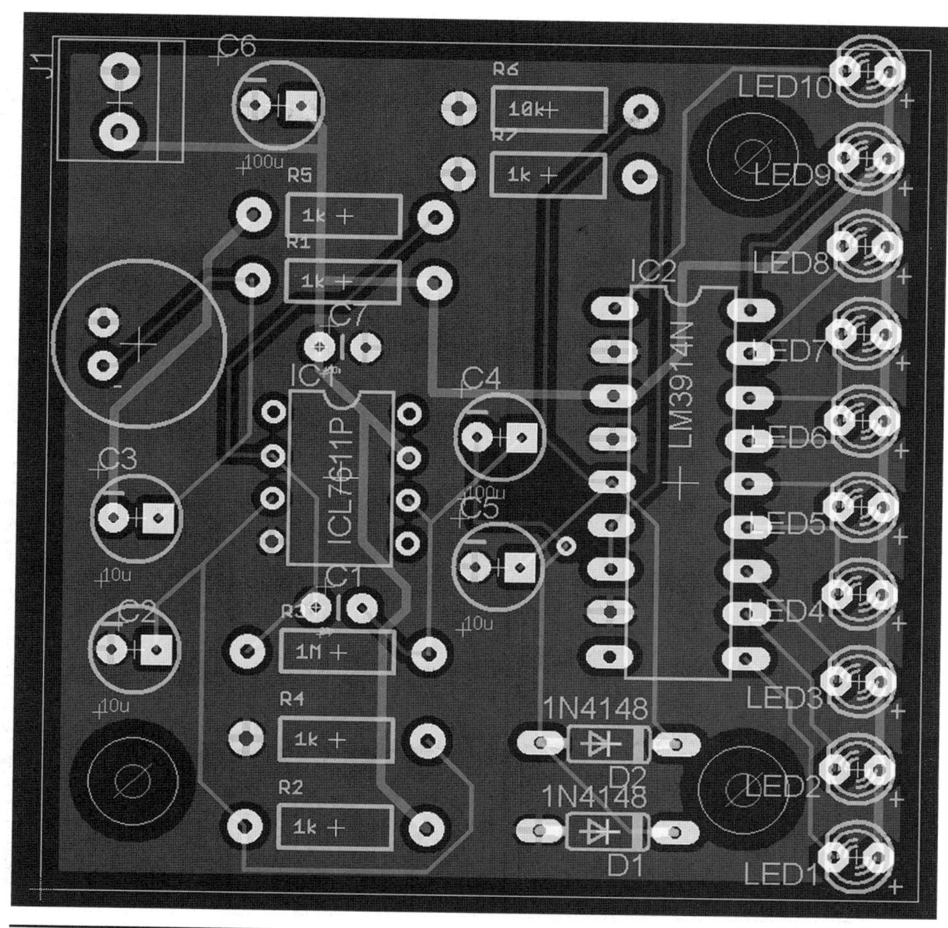

FIGURE 5-21 Fully laid-out board.

Figure 5-22 shows the errors I get. Clicking on one of the errors in the list will highlight the area on the board. In the figure, this is LED10 being too close to the top edge of the board.

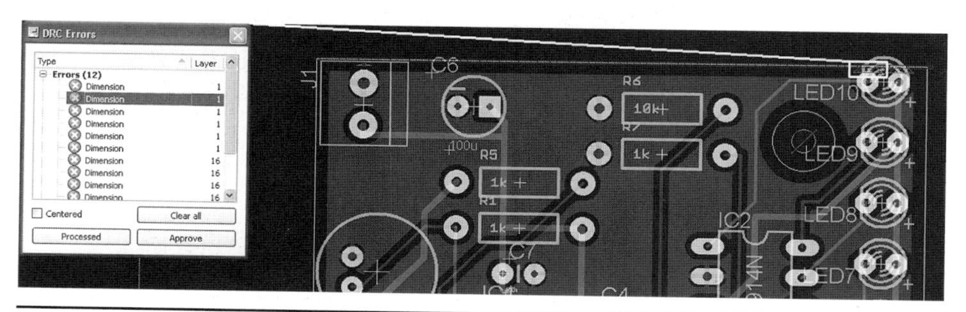

FIGURE 5-22 Results of the DRC.

These errors all relate to things being too close to the top edge of the board. These include

- J1, which is easily moved down a little
- The hole next to LED2, which is too close to the diodes (I solved this by moving the diodes a little to the left.)
- LED10, which is less easy to fix

Because all the LEDs are on a 0.05-in. grid, if we move them all down one grid position, then we will just move the problem to the bottom edge of the board. Therefore, change the grid spacing to 0.025 in. (on the View menu). Rip up everything (again), and move the LEDs away from the edges and closer together. After the autorouter has been run again, there are no air wires. Figure 5-23 shows the result.

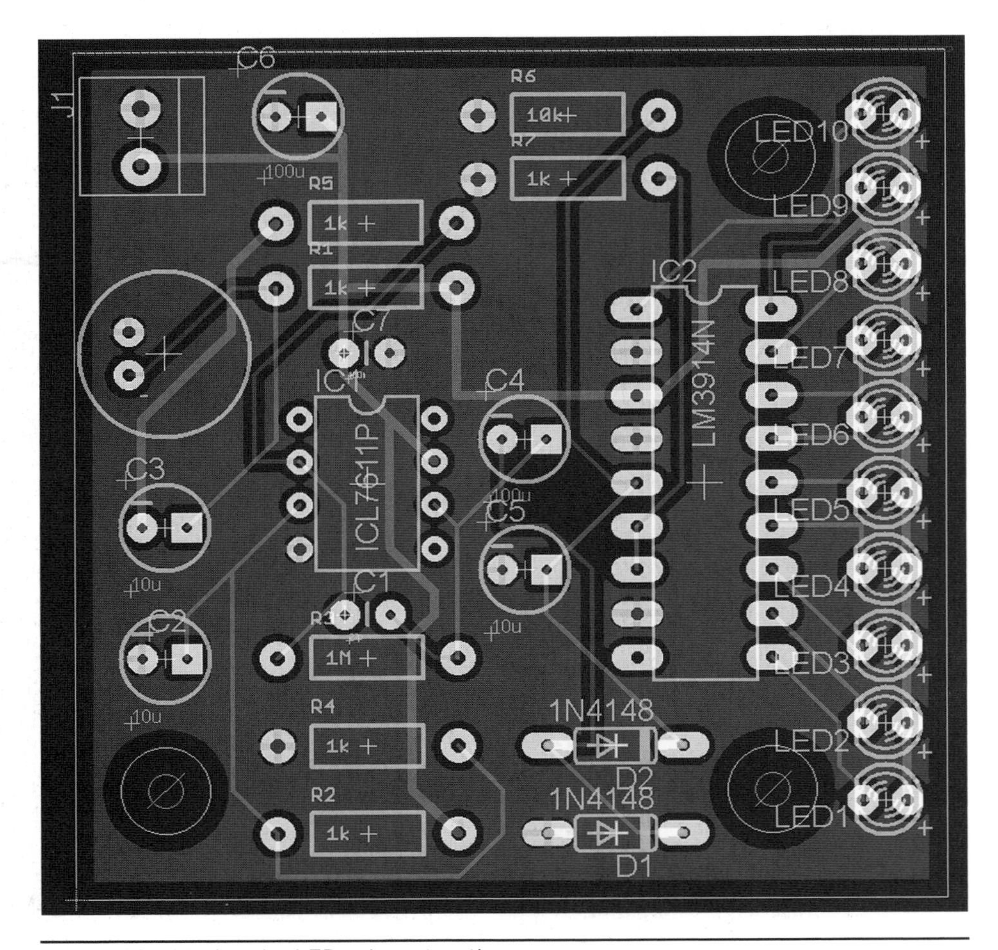

Figure 5-23 Moving the LEDs closer together.

Run the DRC again. You may find an error relating to the ground plane, complaining of "width." You can ignore this.

Let's now move on to the next stage in the preparation of our board—sorting out the silk-screen layer.

Text on the Silk Screen

To make our board look really good, as well as making it easy to assemble and see which component goes where, we need to sort out the writing that will appear on the board, known as the *silk screen*. The "tPlace" is the layer responsible for the silk screen, so let's turn off some of the layers to get a clear view of what we are working on.

Open the Layers list by clicking the Layers command, and select just the layers "Pads," "Dimension," "tPlace," "tOrigins," "tNames," and "tValues." Figure 5-24 shows the result of this.

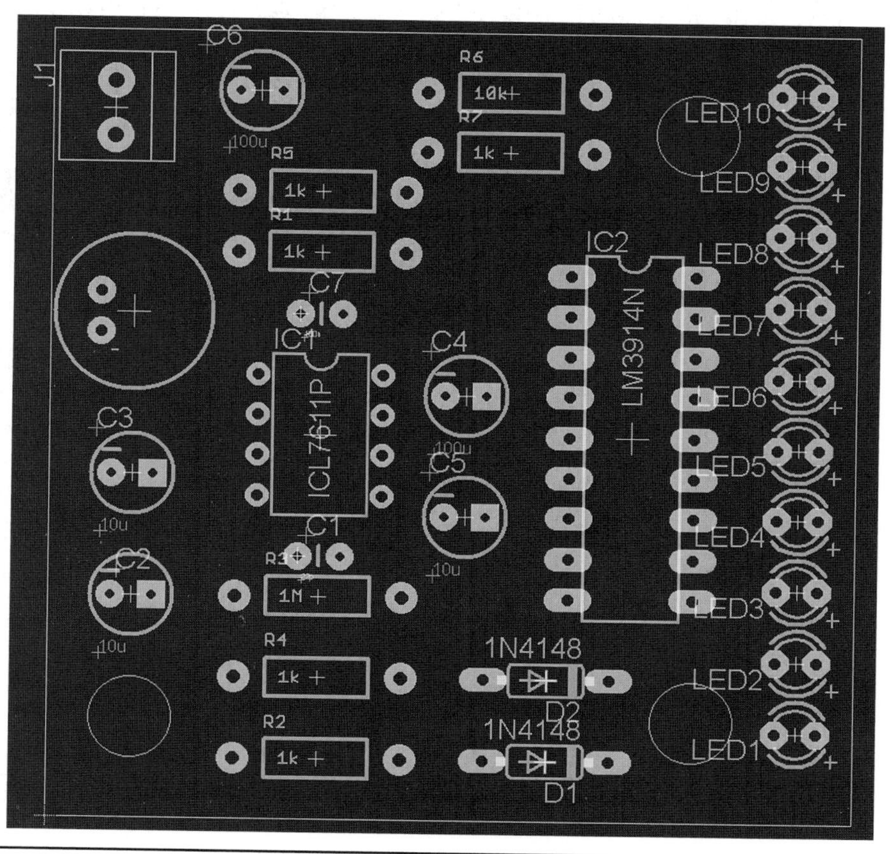

FIGURE 5-24 Hiding layers.

This is actually quite a mess. Labels are in different-sized fonts, and some are overlapped by the part outlines. This is quite common because the parts that we have used have been provided by different manufacturers all with their own standards of EAGLE part design. There is some tidying up to be done.

For instance, there is not really any need to label every LED. Just labeling LED1 and LED10 would be fine. To remove the extra labels, we need to use the Smash command to separate the labels from the parts. In fact, we are probably going to end up modifying, moving, or deleting pretty much every label on the board.

Very few of the labels are in the correct place, so we may as well smash them all by selecting the Smash command, selecting the Group command, dragging over the whole board area, and from the right-click menu selecting "Group: Smash."

Having used Smash to separate the labels from the components themselves, carry out the following actions to do some tidying up:

- Move the labels for LED10 and LED1 away from the holes.
- Delete the labels for LED2 through LED8.
- Delete the 1N4148 value on D1.

The fonts are all different, so we will use the Change tool to change all the labels to be of size 0.05 and font type Vector. Generally, it is better to use the Vector font everywhere on the board. This will guarantee that the text on the actual board will look like the text in the Board Editor.

Click on the Change command (the icon looks like a wrench). Change only changes the last setting that you make in the Change menu. On the menu, you will see "Size." Change this to be 0.05. Select the Group tool, select the whole board, and then right-click and do "Group: Change."

Repeat the process but setting each label to be the Vector font by selecting "Font" and then "Vector" from the Change menu. Then again, set the "Ratio" to be 12.

Finally, repeat the process one last time using the Change tool to set the "Layer" of all the labels and component outlines to be "tPlace" because this is the layer that we will use for the silk screen. Note that this is probably best not done as a Group selection unless you are very careful not to select the board outline, which is in the "Dimension" layer and needs to remain on this layer.

Drag the labels about so that they don't overlap. Note that to move, you must have the Origins layer selected.

When you think everything is changed, go to the Layers menu and deselect "tNames" and "tValue" so that only "tPlace," "Dimension," and "Pads" are selected. The silk screening should now look something like Figure 5-25. The board looks pretty good now.

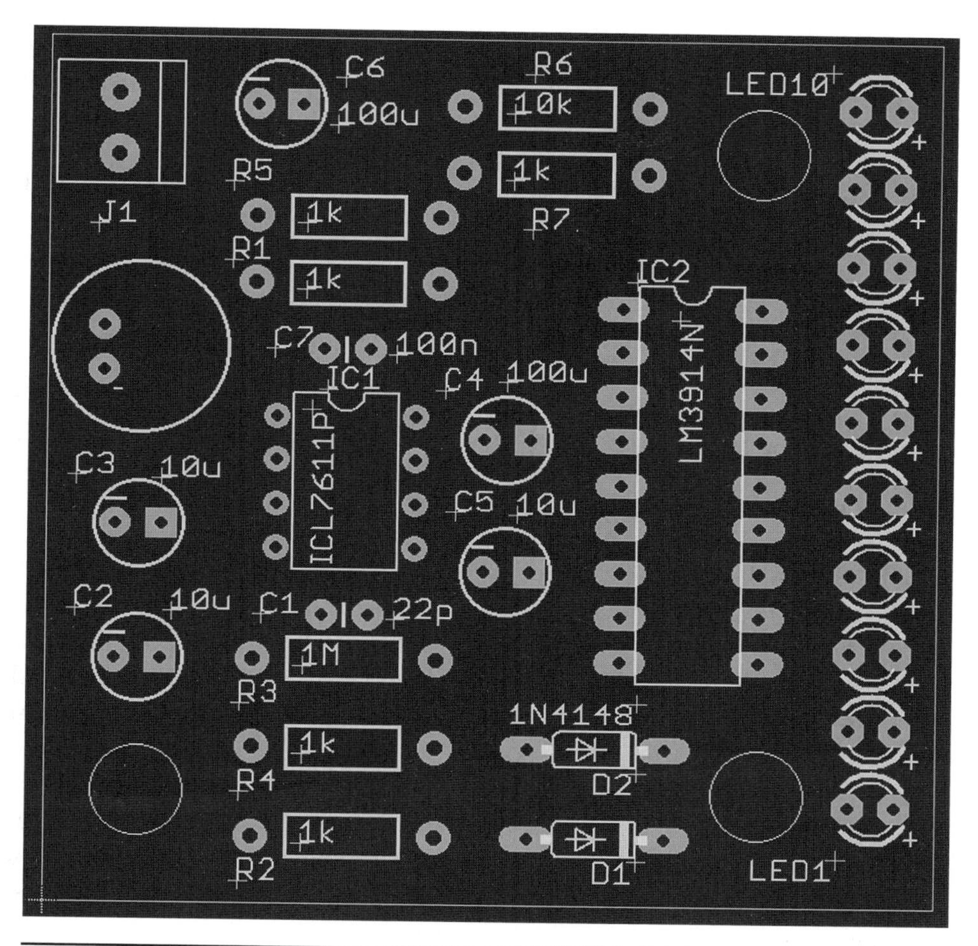

FIGURE 5-25 The "tPlace" layer.

We can just add text to the tPlace layer using the Text command. Use this command to add labels that identify the two terminals of J1. Note that after selecting the Text command and entering the text in the box, you can set the layer ("tPlace") as well as the font size and characteristics to match those we set earlier using the Parameter toolbar. That is,

- Layer: tPlaceGND
- Size: 0.05
- Font: Vector
- Ratio: 12

Figure 5-26 shows the text added.

Figure 5-26 Labeling the terminals of J1.

Add Text to the Bottom

Having legible and well-placed labels will make the board much easier to assemble. However, if you are just making one board for yourself, then it does not matter too much.

The final step in this process is to add some text (a web address) to the silk-screen layer on the bottom of the board ("bPlace"). EAGLE always shows you the view looking down at the top layer, so writing on the underside of the board will be mirrored in appearance. We have a lot of text on the top, so use the Layers command to hide the "tPlace" layer. Then use the Text command to add some text. Remember to select the layer "bPlace" in the Layer dropdown on the Parameters toolbar. You can see the result of this in Figure 5-27.

This completes the through-hole version of the board design. The final version of the board, with "bPlace" hidden, is shown in Figure 5-28. Next, we will look at a surface-mount version of the same design.

Figure 5-27 Adding text to the "bPlace" layer.

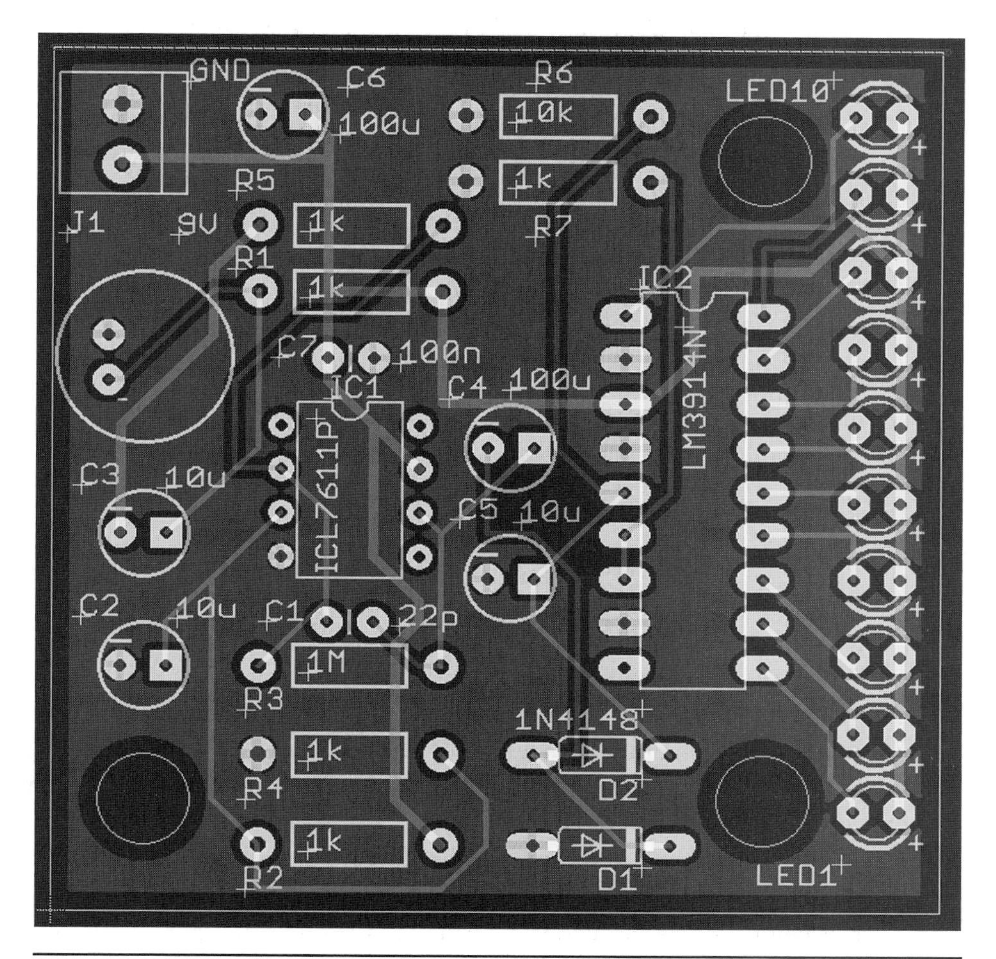

FIGURE 5-28 Final through-hole board layout.

Sound Meter Layout (Surface-Mount)

Having successfully made the through-hole version of the PCB, it is time to look at producing a second, more modern version of the board that uses surface-mount components.

In an ideal world, you would be able to use the same schematic but with surface-mount device (SMD) package variants of the components. Unfortunately, this is not possible in EAGLE. This does not mean, however, that you have to start the whole design over. The process is to take a copy of the schematic for the through-hole version and then go through all the components, replacing the through-hole parts with SMD equivalents.

Create a New Schematic and Board

Make sure that you have saved your through-hole schematic. You are now going to save a copy of the schematic under a different name but in the same project as the original. Thus, with the Schematic Editor open, select the option "Save As…" Then save the schematic with the name `soundMeter_sm.sch`, as shown in Figure 5-29. The addition of `_sm` indicates that this is the surface-mount version. It has no special significance as far as EAGLE is concerned; it just allows us to distinguish the two versions.

After saving the copy of the schematic, you will notice that the Title field in the schematic's frame will now also be updated to `soundMeter_sm`, allowing us to tell our schematics apart should we print them out.

Saving a copy of the schematic in this way will automatically create a copy of the board called `soundMeter_sm.brd` that is linked to the new schematic.

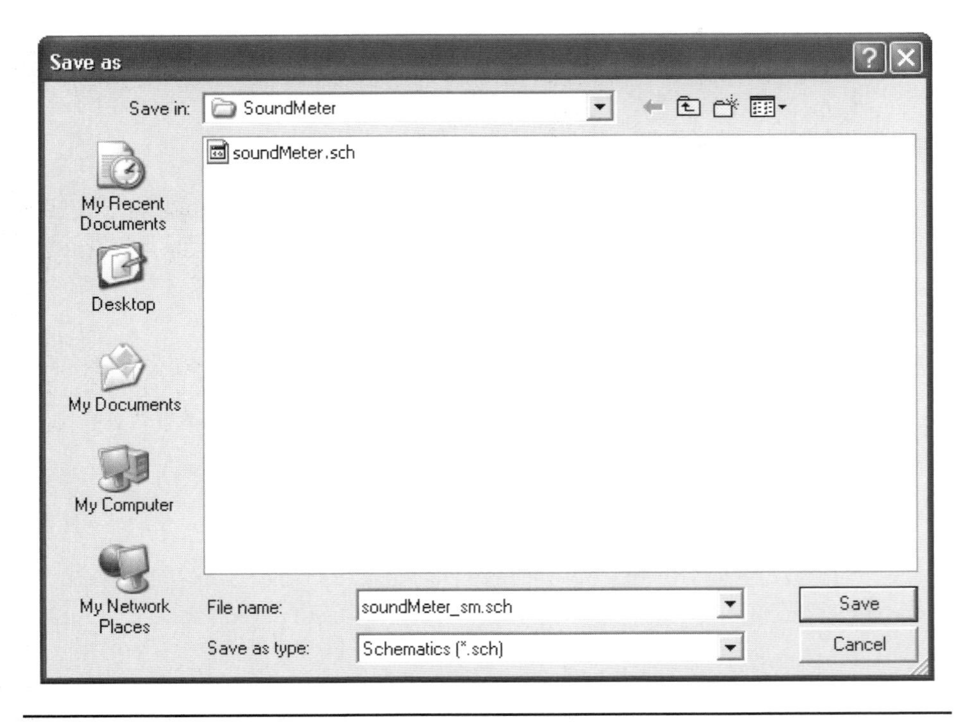

Figure 5-29 Saving the schematic to copy it.

Swap Parts on the Schematic

The surface-mount schematic (`soundMeter_sm.sch`) is currently exactly the same as the through-hole schematic. We now need to go through the parts on the schematic and swap them for equivalent SMD packages. The microphone and screw terminals will remain as through-hole devices.

Let's start with IC1. Click the Replace command on the Schematic Editor's Command menu and then use "*7611*" as the search string (Figure 5-30).

Looking at the search results, you can see that the chip is available as DIL08 (through hole) and SO08 (small-outline eight-pin SMT). So select the SMT option, and hit "OK." Now click on IC1 on the schematic. It may seem like nothing has happened, but if you now look at the board layout, you will see that IC2 now looks very different (Figure 5-31).

You can also see that the routing is now, of course, all in a mess. This is only going to get worse, so we may as well go back to air wires by doing a group ripup. To do this, click on the Ripup command, then click on the Group command, drag over the whole area of the board, and right-click and slect the option "Ripup: Group."

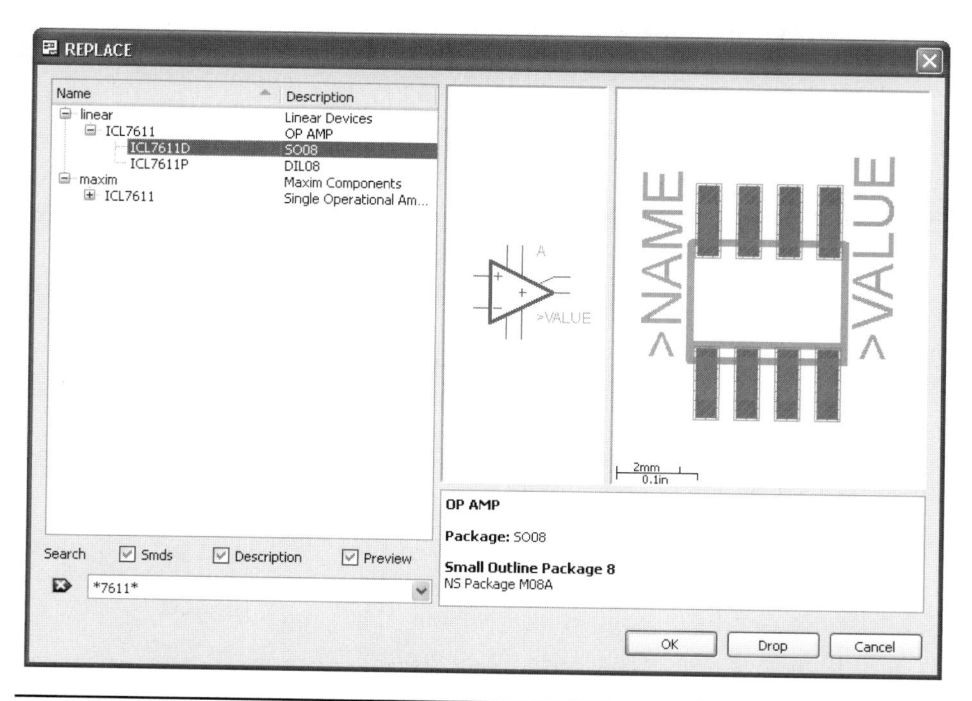

FIGURE 5-30 Replacing IC1.

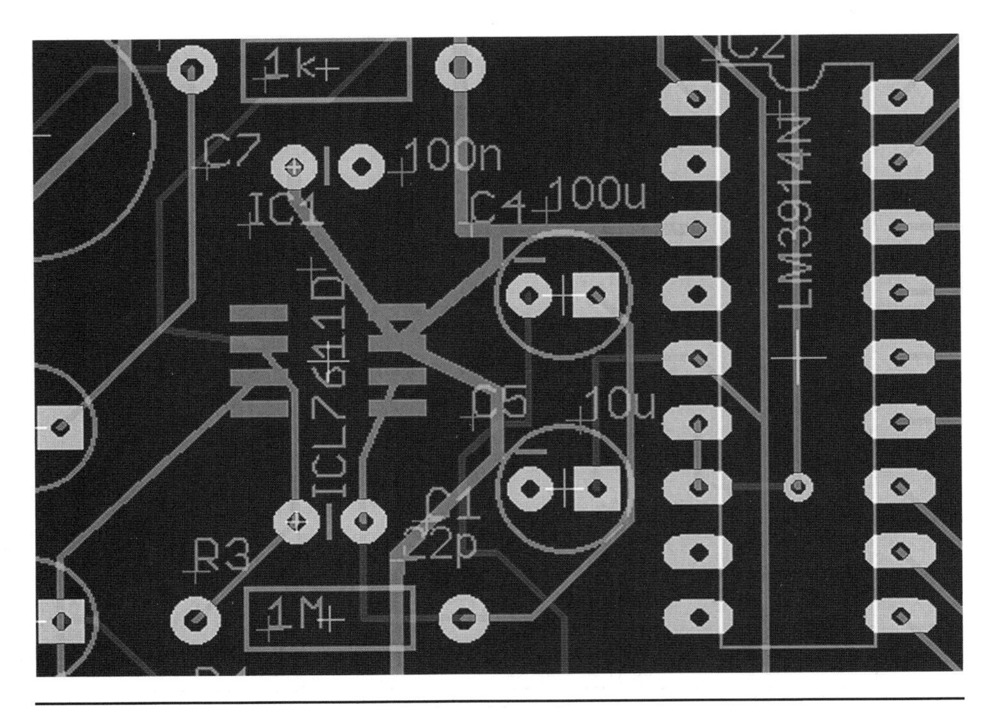

FIGURE 5-31 IC1 SMD package on the board layout.

We may as well set the Board Editor back to a useful state ready for layout by just showing the layers "Top," "Bottom," "Pads," "Vias," "Unrouted," "Dimension," "tPlace," and "tOrigin." We were kind of lucky with IC1 in that there was a drop-in SMD replacement using the same pin names. Unfortunately, things are not quite so easy for IC2. Using the Replace tool and a search string of "*3914V*" will find an SMD equivalent of the chip, but when we try to do the replacement, we will get an error message `Date A in the old version of device set LM3914N can't be mapped to any gate in the new version of this device set (neither by name nor coordinates).`

This is so because the part has been designed differently from the original and uses different names for many of the pins. For example, the connections to the LEDs are called LED1, LED2, and so on rather than L1, L2, and so on (Figure 5-32).

In Chapter 11 you will see how to copy and modify parts from libraries, but for now, we will reconcile ourselves to having to manually replace IC2. This may seem a little daunting but is pretty straightforward. A good tip is to print or take a screen capture of the existing schematic to use as a reference before deleting IC2 prior to replacing it.

Use the Delete command to remove IC2 from the schematic. This will leave a big hole in the schematic but otherwise leave the net lines in tact. Unfortunately,

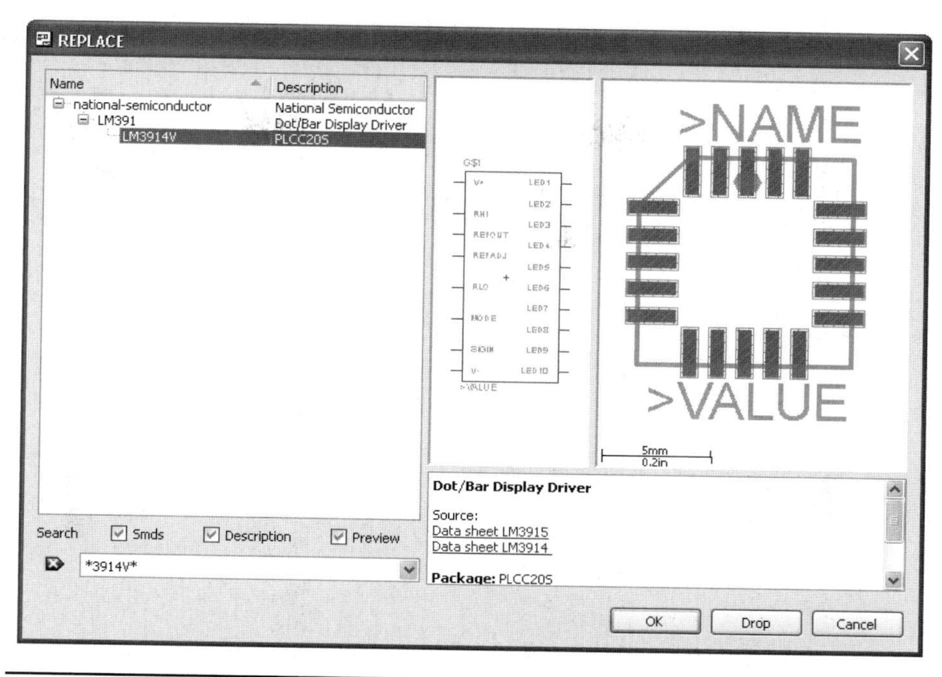

FIGURE 5-32 SMD package for IC2.

these will also be in the wrong positions when we add in the SMD version of IC2, so make a Group delete of all the net lines close to where the new IC2 will be added. Now use the Add command to add in the new version of IC2 (search for "*3914V*"). The result will look something like Figure 5-33.

Using your screen capture or printout of the old schematic as a reference, connect everything up again so that the schematic looks like Figure 5-34. Note that the newly added IC2 will need to be dragged onto the board because it will initially be placed outside its outline.

You will also need to copy a V+ terminal and connect it to the V+ pin of IC2. Switch back to the schematic view, and replace the components as per Table 5-2.

TABLE 5-2 SMD Replacements

Part	Search for
R1 to R7	RESISTOR1206
C1, C7	CAP1206
C2, C3, C5	CAP_POLC
C4, C6	CAP_POLD
D1, D2	DIODESOD
LED1 to LED10	LED1206

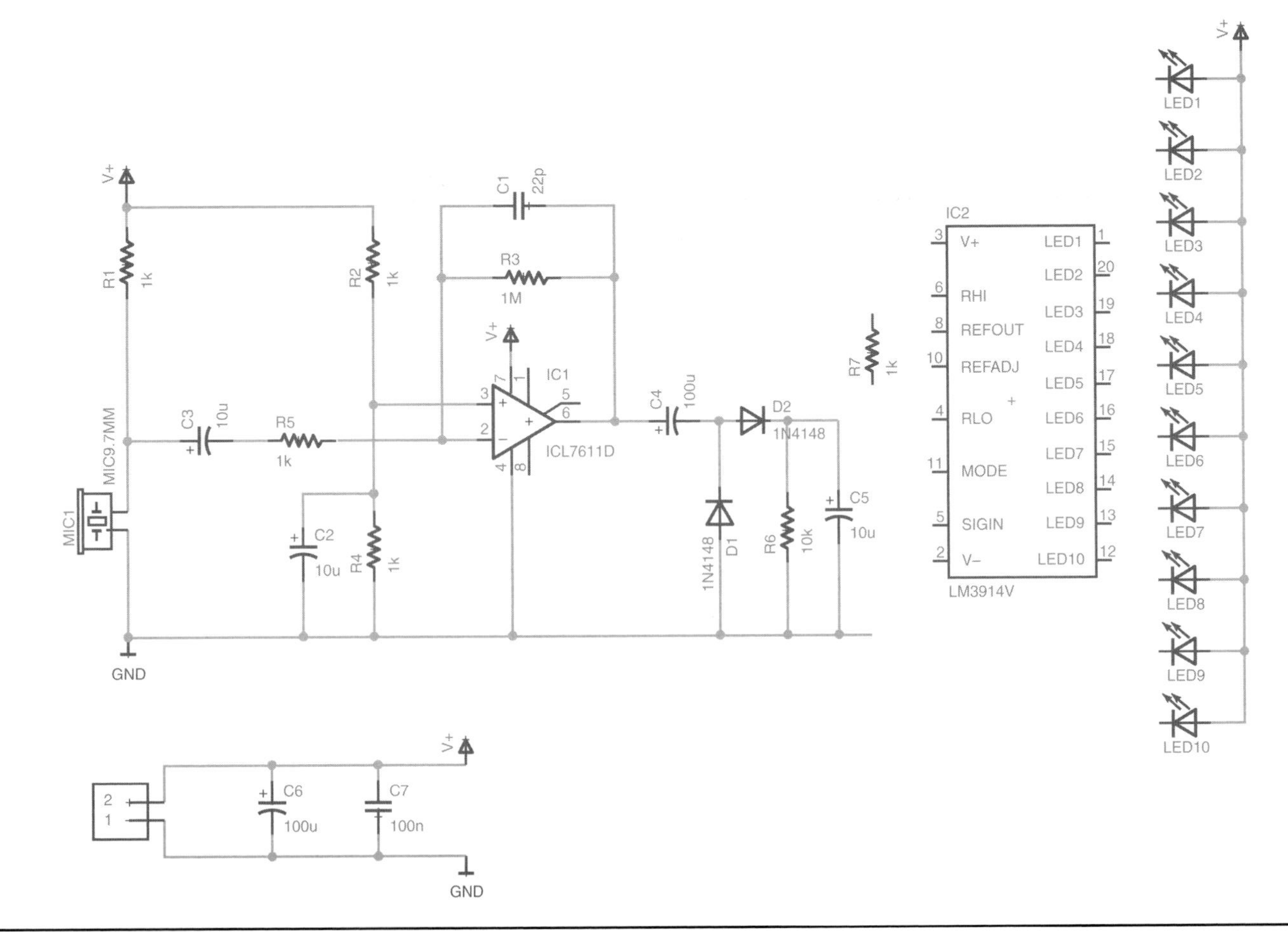

FIGURE 5-33 Making room for the replacement IC2.

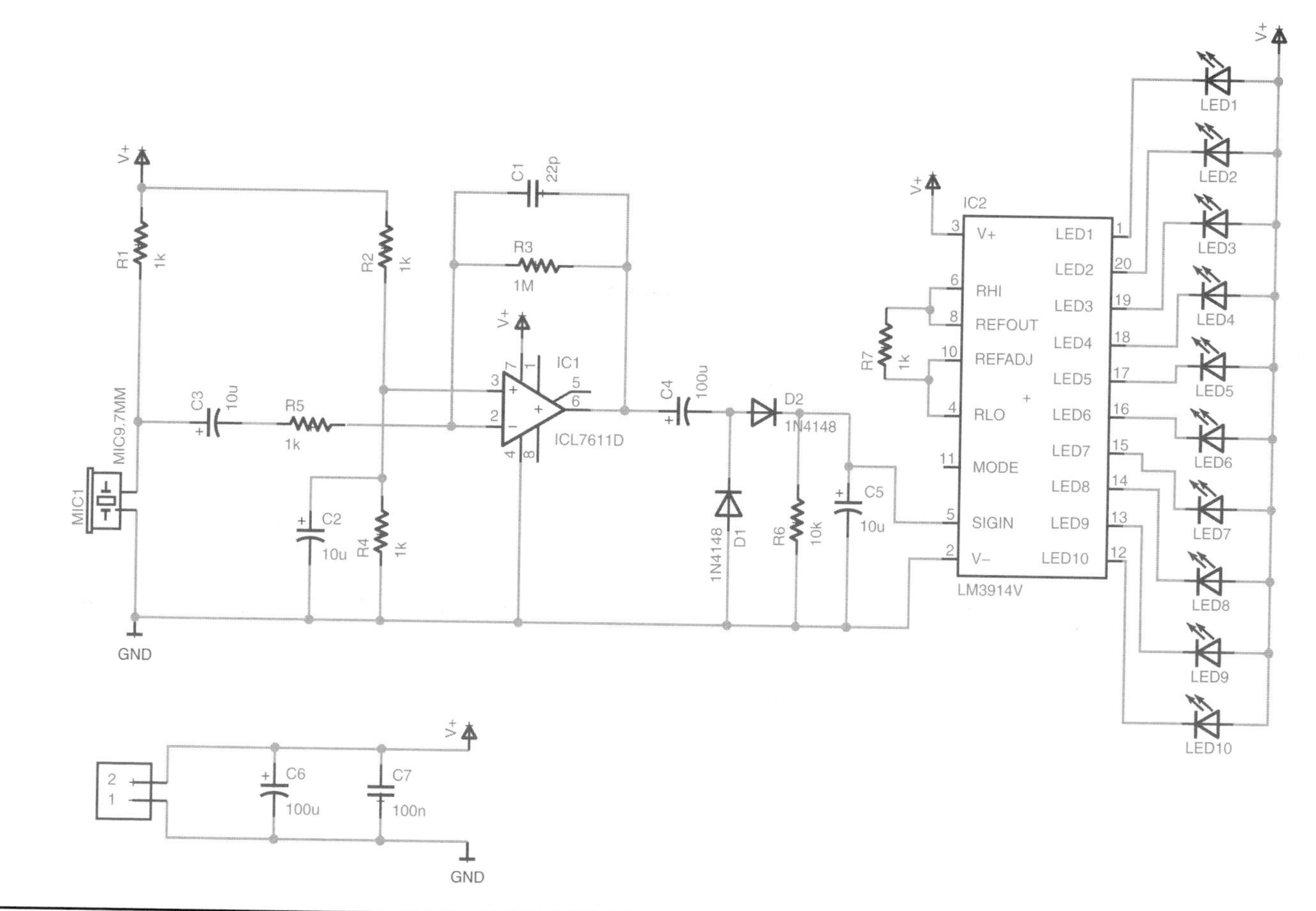

FIGURE 5-34 Revised Schematic with IC2 replaced.

As you replace the components with their SMD equivalents, switch over to the board layout every now and then and notice how the components have changed and how much more sparse the board is looking.

Now that we have all the components swapped over to SMD equivalents, we need to do a bit of tidying up and reroute the board.

Resize and Tidy the Board

You can pack a lot more into an SMD design than a through-hole design. So we can take the opportunity to make the board a bit smaller and move the LEDs closer together. The mounting holes on the first design were also rather large, so we can reduce them to just two smaller holes. Note that to be able to delete or move holes, you will need to make the "Holes" layer visible.

It is easier to rearrange the components first and then adjust the size of the board to just contain them. Rearrange the components on the board until you are happy with the result, and then run the autorouter again. When running the autorouter this time, change the routing grid to 10 mils when the Autorouting window opens. This will allow the router to work around the smaller pitched pins of the SMD components. The final result might look something like Figure 5-35.

Manual Layout

In the example layouts thus far, we have made use of the services of the autorouter. It does a pretty good job and is certainly much quicker than laying out the board by hand. However, it is a good idea to know how to lay out a board manually for a number of reasons:

- The autorouter cannot always completely lay out a board that a human mind can.
- For some circuits where the designer (that's you) wants to keep certain track lengths short and thick, a full manual layout usually will yield better results.
- It's fun. If you like puzzles, then you will probably enjoy laying out a board by hand. It can be a fascinating challenge to lay out a board so that all the tracks are as short and direct as possible with a minimum number of vias. This is especially true when it's a board that the autorouter believes to be impossible—it's you against the machine!

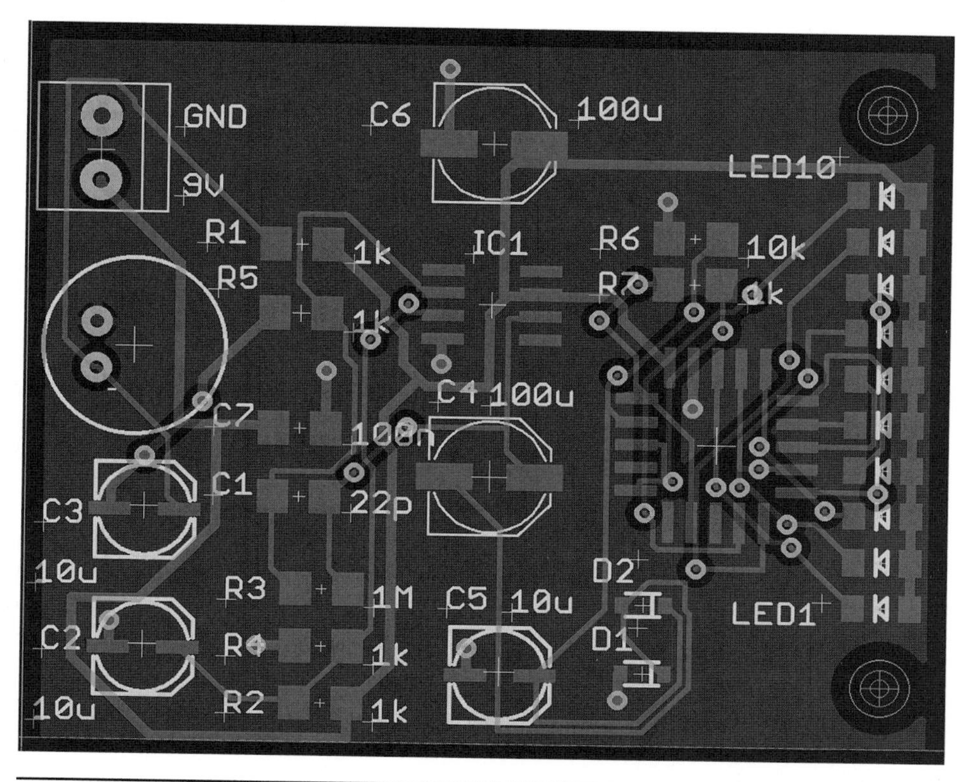

FIGURE 5-35 Final SMD board layout.

I am going to lead you through laying out the through-hole version of the sound meter project starting with the same component positions as we used in the automated layout. So that both layouts can be available to compare, you will find the design files for this called `soundMeter_manual.sch` and `soundMeter_manual.brd`. The staring point is the unrouted board, as shown in Figure 5-36.

As with the automated layout, we have kept the ground plane on the bottom layer.

I like to start a manual layout with the power-supply nets. In this design, many of these are provided by the ground plane, so we can start by laying out the positive supply.

Once the Route tool has been selected, the Parameter toolbar will display the options available to us. I described these options at the start of this chapter. In Figure 5-37, I have set these parameters ready for routing the positive power line on the top layer.

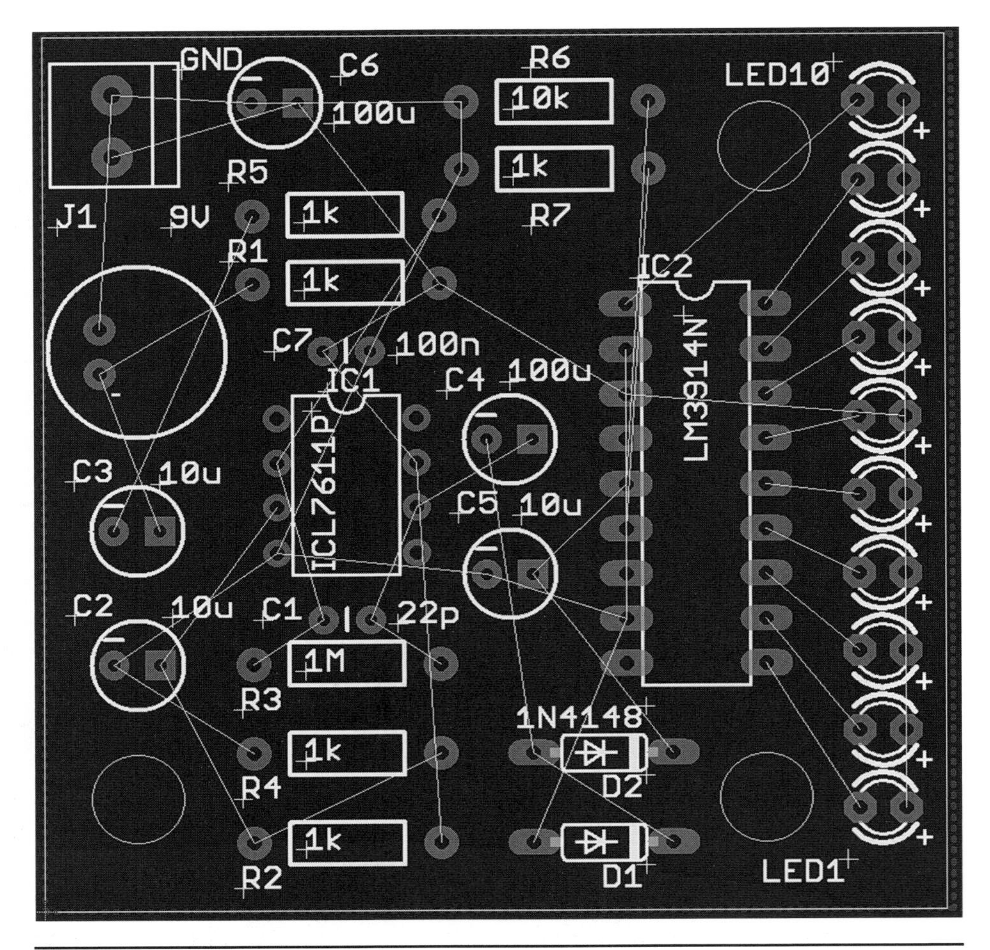

FIGURE 5-36 Unrouted board.

FIGURE 5-37 Routing toolbar.

The first thing to note is that the Layer dropdown is set to "1 Top." We then have to select the fourth of the "Wire bend" types. It does not really matter which one you chose, but I find that this one produces nice-looking layouts. The miter radius is set to 0 because we do not need the corners of the tracks to be rounded. The width is set to 0.02 because this is the track width that we specified for the power-supply net class.

We have also opted to use square vias with the size set automatically. But we will try to make a layout that does not need any vias.

The only parts of the design that require significant current to flow are the power supply to IC2 and the supply lines to each LED. In fact, the supply to IC2 could be as much as 200 mA if all the LEDs are lit.

When using the Routing tool, start at a pad, and an air wire will become highlighted along with the rest of that air wire's net. This indicates all the possible target points for your routing, so just move the cursor along the route you want to take, left-clicking to make a waypoint in the route. When you arrive at a valid destination, the route will stop automatically, and you can go off and find another air wire to route.

If you go wrong, you can, of course, use the Undo command. If you find an earlier mistake that you want to correct, the just select the Ripup command, and click on the tracks that you want to route again and they will revert to being air wires.

You can also modify the path of the routing using the Move command if the track is just a little off and does not need complete rerouting.

If you are concerned about just how wide your tracks need to be, then a calculator such as the one at http://circuitcalculator.com/wordpress/2006/01/31/pcb-trace-width-calculator/ can be very helpful. Using this calculator, for a typical PCB with 1 oz/ft^2 of copper and a track 1 in. long and 20 mil wide will only increase in temperature by 10°C when a current as large as 1.5 A is flowing through it.

Thus the tracks on our circuit should barely get warm at all with the currents we are using. For high-current circuits, though, overheating of the tracks eventually will damage the circuit board and cause it to fail because, ultimately, the copper will melt.

Figure 5-38 shows the positive supply to both ICs routed on the top layer. Notice how we have kept the route of the positive supply from J1 to IC2 as direct as possible. We have also routed the supply to the LEDs around the edge of the board so that the long track does not cut off a section of the board near the top, which would then be difficult to route around.

My next step is normally to route easy wins, that is, things such as the connections between IC2 and the LEDs. Figure 5-39 shows these routed on the bottom layer. Remember to change the layer in the dropdown list to bottom and to change the width to 0.01 (10 mils).

At first, it will seem like the routing on the bottom layer is interfering with the ground plane. When you press "Ratsnest," gaps around the tracks will magically open up.

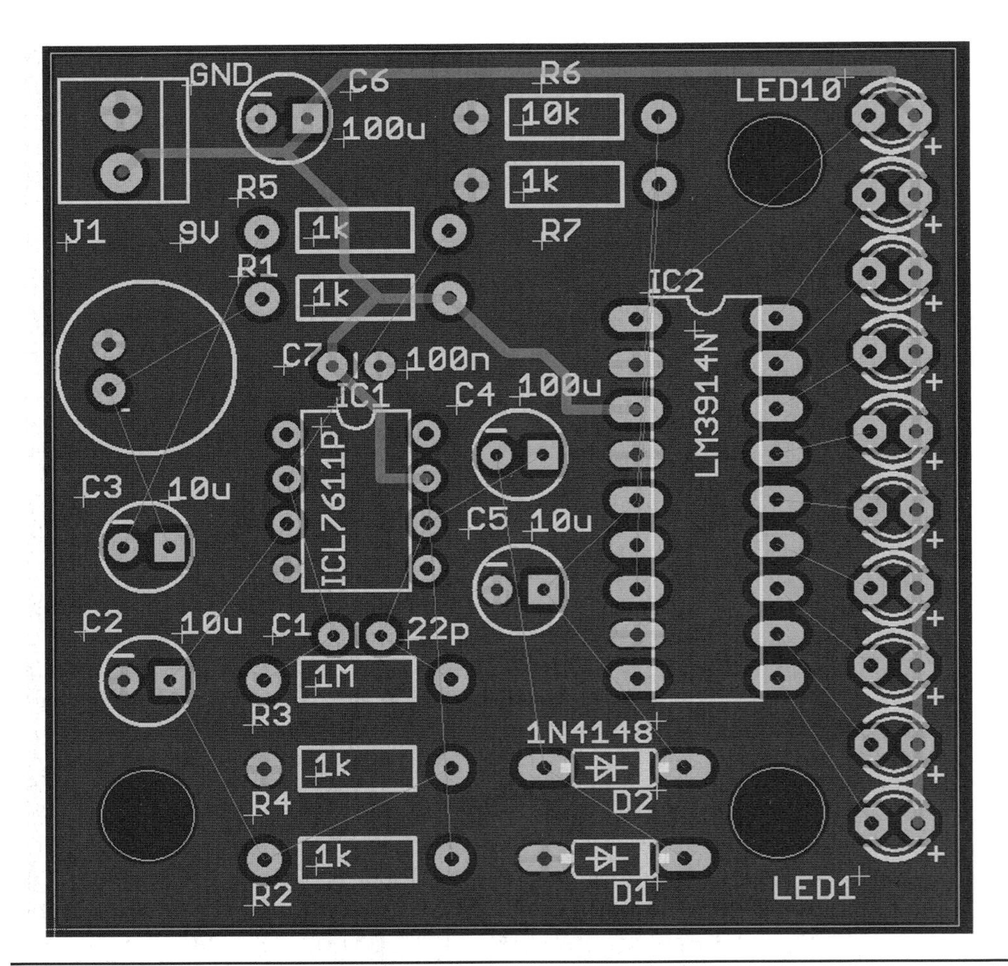

FIGURE 5-38 Routing the supply nets.

There are some other obvious tracks to be laid around the center of the board. In particular, we want the tracks for the feedback components around IC1 (R3 and C1) to be kept short and close to IC1. While we are routing on the top layer, let's route anything we can do without to take a long route around the board. So let's complete them using the top layer (Figure 5-40).

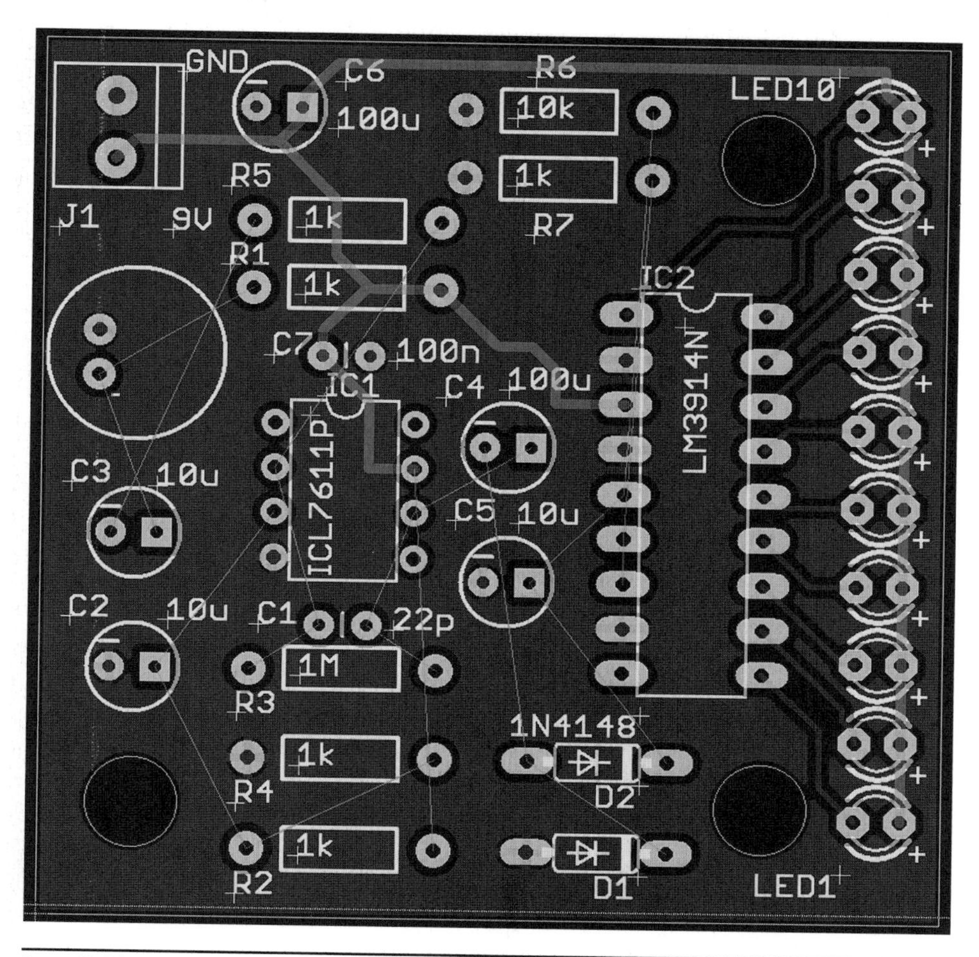

Figure 5-39 Routing the LEDs.

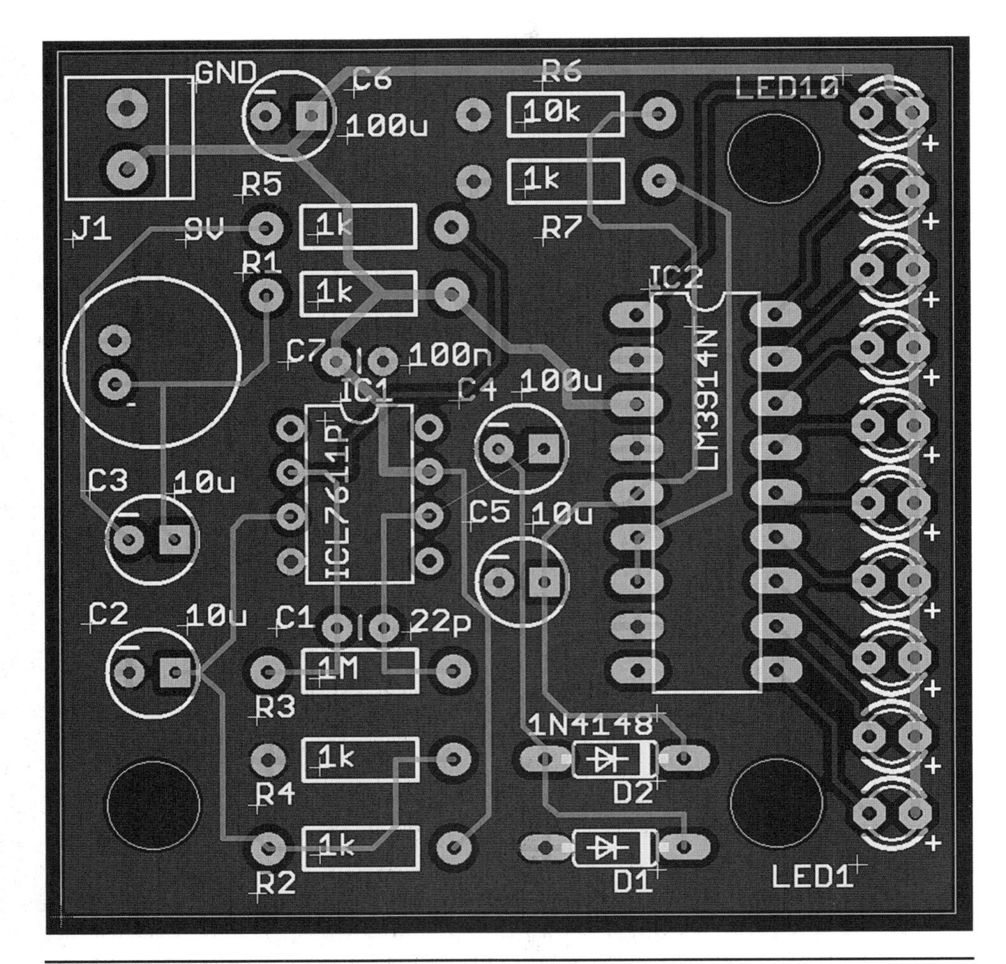

Figure 5-40 Routing the rest of the top layer.

We now have just one air wire (between IC1 and C4) that cannot be easily routed on the top layer. Therefore, switching to the bottom layer, we can add this in. The final result of the manual routing is shown in Figure 5-41. You might like to compare this with the automatic routing shown back in Figure 5-28.

Summary

In this chapter, we have designed two versions of the same board, one using through-hole components and one using surface-mount technology. Having produced our board design, the next logical step is to generate the design files that will allow the boards to actually be made. This is the topic of Chapter 6.

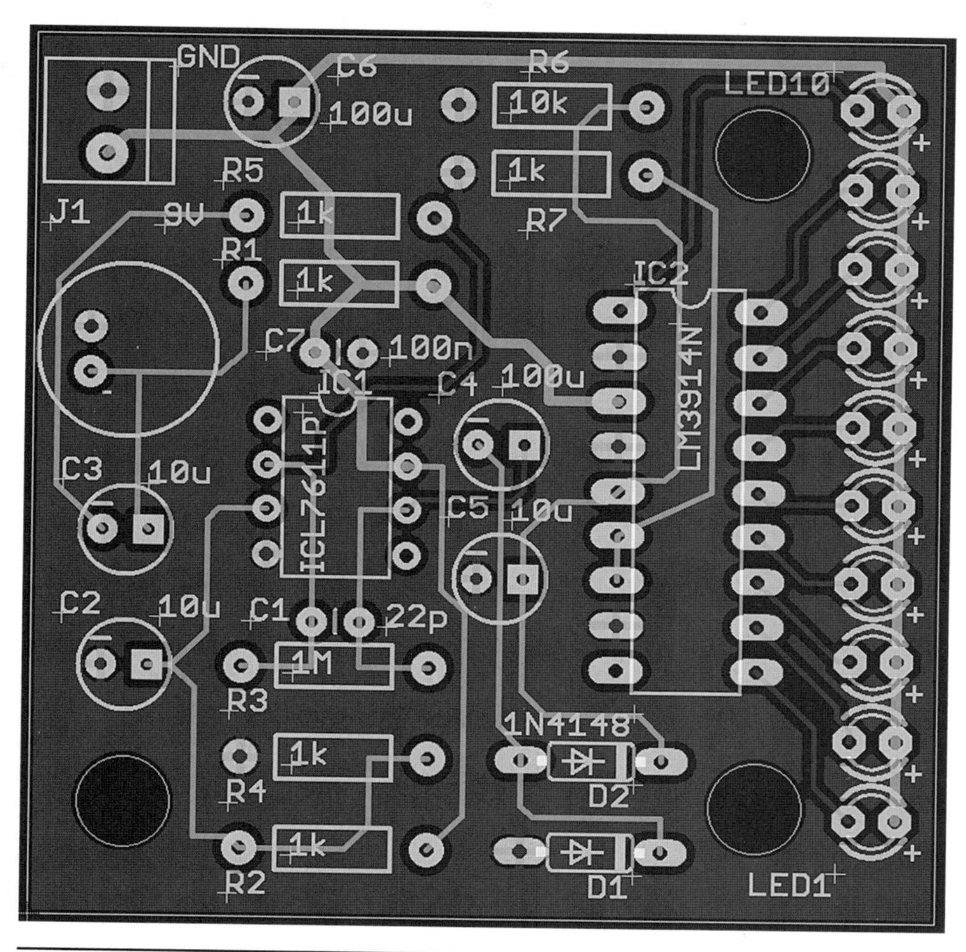

FIGURE 5-41 Final manual layout.

CHAPTER 6

Printed Circuit Board Fabrication

At the end of Chapter 5, we had a finished board layout. We now need to use that to generate the files that can be sent to a PCB fabrication company that will in due course send us back some real PCBs.

The CAM processor is the EAGLE module that converts the layers in the EAGLE PCB design and produces the industry standard files required to fabricate the PCBs. This chapter explains this process and shows you how to produce such files using the example sound meter project developed in Chapters 4 and 5.

Although you can create your own PCBs using photosensitive PCBs and photo etching, it can be a messy and tricky business. It is also relatively expensive because the developer and etchant chemicals do not last. They are also toxic and difficult to dispose off because you cannot simply wash them down the drain. Because you can have a double-sided silk-screened drilled PCB made by a fabrication service for a dollar or two per board, even for runs of 10 or fewer boards, unless you are in a real hurry or just enjoy the process, there is little point in making the boards yourself.

Having said that, later in this chapter we will look at photoetching your own PCBs.

Gerber Files

Although some PCB fabrication services will accept EAGLE design files directly, most require you to produce a set of what are called *Gerber files*. These files are an industry standard for PCB fabrication. You will normally be required to produce seven files.

Table 6-1 Gerber File Types for CAM

Myboard.GTL	Top layer (copper)
Myboard.GBL	Bottom layer (copper)
Myboard.GTS	Solder stop mark (top)
Myboard.GBS	Solder stop mark (bottom)
Myboard.GTO	Silk screen (top)
Myboard.GBO	Silk screen (bottom)
Myboard.TXT	Drill

Table 6-1 shows the files that you should submit. Each of the files has a different extension that indicates its contents.

Loading a CAM Job

The process of generating the CAM files is extremely configurable. You can essentially use any of the layers as the basis for any of the preceding Gerber files. This flexibility is very powerful but also allows plenty of room for error if you are designing your own CAM configuration from scratch.

By far the easiest way of generating the files is to use a CAM configuration (called a *CAM job*) that has been designed by someone else. In this book, we will use the excellent CAM job designed by Sparkfun. This is available free of charge from Sparkfun at https://github.com/sparkfun/SparkFun_Eagle_Settings/tree/master/cam.

To install this into EAGLE, download the file `sfe-gerb274x.cam`, and copy it into the `.cam` folder in your EAGLE installation directory.

Running a CAM Job

To load the Sparkfun CAM job, select the "CAM Processor" option from the File menu while you have the board you want to create Gerber files for open. This will open the default CAM processor job, as shown in Figure 6-1.

To open the Sparkfun CAM job, select the option "Job..." from the "Open" option of the File menu on the CAM Processor window. Then select `sfe-gerb274x.cam` from the list, and the Sparkfun CAM Job will open up (Figure 6-2).

As you can see, the difference between the default CAM job and the Sparkfun one is that the Sparkfun CAM job has a row of tabs, one to generate each of the Gerber files. For example, the one shown in Figure 6-2 is labeled "Top Copper."

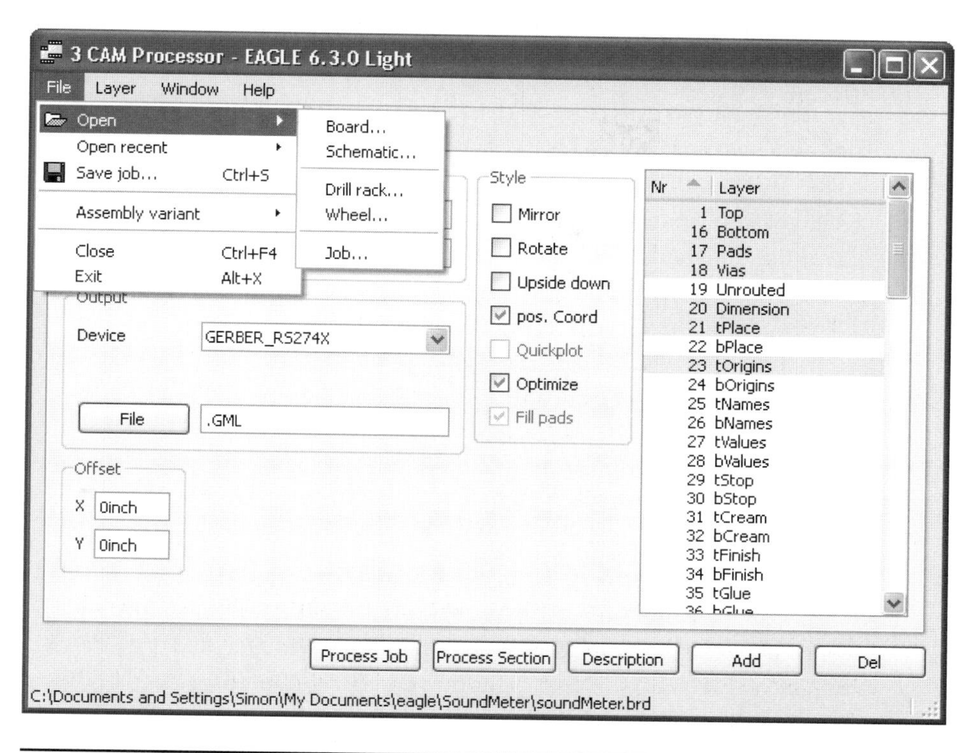

FIGURE 6-1 Default CAM job.

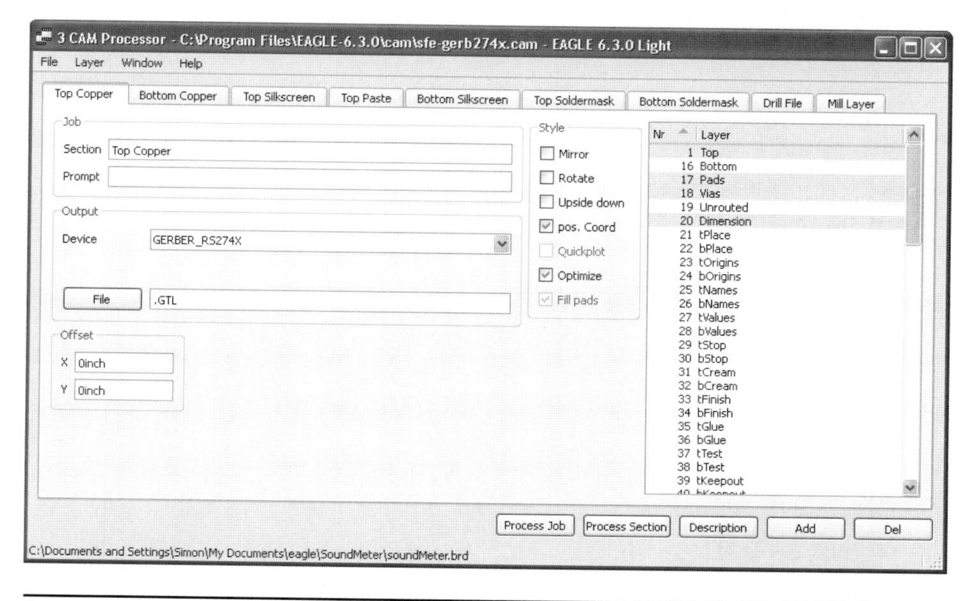

FIGURE 6-2 Sparkfun CAM job.

As you might expect, this is responsible for generating the top copper Gerber layer (`.GTL`).

In the "Output" section of the tab, you can see the "Device" field with a dropdown list next to it. This can be set to a number of other types of CAM format. This should be left as `GERBER_RS274X`.

On the right-hand side of the tab is a list of layers, with some of the layers highlighted. The highlighted layers are the ones that will become copper when the CAM job is processed. The other tabs each work in a similar way.

To generate the Gerber files, all we need to do is to hit the "Process Job" button. The files will be generated and placed in the project folder. Figure 6-3 shows the full set of generated files.

Measure Twice, Cut Once

At this point, it is extremely tempting to send the design files off to the fabrication service. However, the old carpentry maxim, "Measure Twice, Cut Once," is very relevant here. There is nothing worse than submitting a job to be made only to suddenly realize that you had forgotten to check something and you would just have to pay for and await the return of a set of useless boards. Having said that, if the boards are just for a prototype, then a little surgery on incorrect boards is often possible, cutting a track here and soldering a link there.

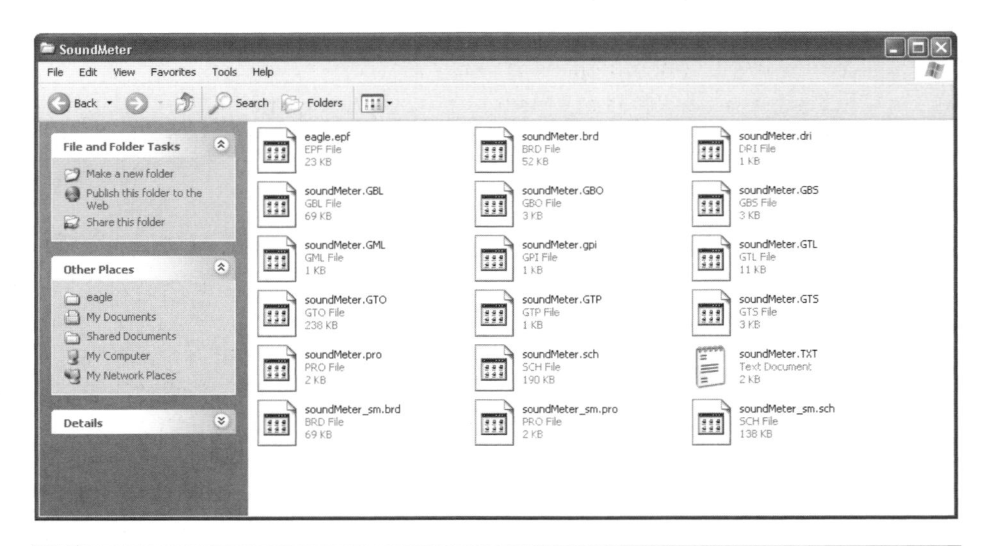

FIGURE 6-3 Generated Gerber files.

So now is the time to check your design once more to make sure that both the electric rule checker (ERC) and the design rule checker (DRC) have been run. Although these will catch a lot of problems, they will not guard against a design that is simply faulty.

To illustrate this with a real problem, in my original design, I had D1 the wrong way around throughout the design, even in the schematic. I didn't actually catch this problem until the boards came back. However, this was easily remedied by inserting the diode the "wrong" way around on the PCB. I then had to retrace my steps and redo all the PCB designs.

Another problem is that it is often difficult to know if the parts you have picked out of the library have exactly the same package and pin dimensions as the components that you have. If this is the case, then it is well worth making a paper prototype. To do this, simply print out the board layout and try poking the component leads through the holes to make sure that everything fits. This will also highlight any problems with the third dimension (height) that you never see using EAGLE. For example, it may become apparent that one component is sticking up too much. For example, the SMD version of the sound meter project probably could benefit from having all the LEDs on the bottom of the board so that there are no components sticking up above the LEDs, allowing them to be mounted flush against a window on whatever box the board is to be housed in. It's surprising what a difference it makes having something concrete to handle.

Submitting a Job to a PCB Service

Finally, it's time to find a PCB service and send off the design files.

PCB services aimed at the maker are an ever-expanding and changing area. Therefore, before selecting a service, do some research. The main things that you need to consider are

- **Cost.** How much will it cost you for your project. If you are just making a project for yourself, then you may only want one board. Wasteful though it may be, you may find that you can get 10 boards from one supplier for less than the cost of one board from another supplier.
- **Speed.** How long will it take for the boards to come back after you have sent over the Gerber files? Wherever possible, look for information on the electronic forums about the actual turnaround time.
- **Quality.** These days you would be unlucky to receive a low-quality board. These things are made by high-quality machines, and there is little practical to go wrong.

- **Design rules.** Each service will have its own design rules. Sometimes these are available as a download for EAGLE, but I would use a more universal set of design rules such as those of Sparkfun and simply check that your track thicknesses and spacings are greater than those specified for the service. Generally, they will be.

These items tend to be a tradeoff, so if you want the boards fast, they are unlikely to be low cost. The size of the board also often makes a big difference. Some services simply charge by the square inch, and others have certain cutoff sizes, so if you stay within certain dimensions, the boards are much cheaper.

Most of the services you find for prototyping and small-batch numbers will be so-called batch PCB services. These operate by collecting together groups of PCB designs from lots of customers and combining them into a single order. This requires the service to wait until it has a sufficient number of boards to make it worth making a large PCB panel containing all the individual designs that are cut away from each other during manufacture. This means that the delay can be very variable, and you may get your boards really quickly or it may take weeks. Look for maximum and minimum service times. Also look at the bigger services in this area, such as OSH Park, Itead Studio, and Seeed Studio.

Follow the Instructions

Each service will have a detailed set of instructions on how to use their service. Frankly, if they don't have this, then you probably should steer clear anyway. For example, we will look at the instructions provided by IteadStudio's Open PCB service. You can find the instructions for the company's 5- by 5-cm basic green PCB service at http://imall.iteadstudio.com/open-pcb/pcb-prototyping/im120418001.html.

In the section "Requirements on the Design and Gerber Files," you will find instructions on the dimensions of the board, which can be an unusual shape but must fit within the square specified. It also tells you the Gerber files they require.

Most important, it also tells you the minimum line width and text height for silk screens of 6 and 32 mils, respectively. You would need pretty good eyesight to see text that small.

As for the copper, you will see that instructions specify a recommended width and separation of 8 mils. Because our thinnest tracks are 10 mils, we should be just fine.

It is also worth noting that if you want to make a tiny PCB, it may be a little irksome that because your design is at the minimum board size, it may be tempting to do something called *panelizing*. This involves putting multiple copies of the board design onto a single PCB design, with a row of very close together holes separating the individual boards so that they can be separated by snapping them off. This is usually not allowed, although if you simply mark a line to cut on the silk-screen layer, the service will not object.

You also may be given the choice of different board thicknesses and finishes. Very small boards can be thin: 1 mm typically would be fine for a board smaller than 50 mm^2. For bigger boards, 1.6 mm is a common thickness. There also may be options for being lead-free hot-air solder leveling (HASL) and restriction of hazardous substances (RoHS). HASL will make the boards easier to solder but is by no means essential. RoHS is a European Union directive intended to improve the environmental impact of electronic manufacture. If you plan to sell your PCBs in Europe, you should conform to this option.

Depending on the service, you will either need to upload the files, possibly enclosed in a zip file, or separate and associate with an order number, which may require you to rename the files to include your order code. Follow the instructions carefully.

The next step is that the files will undergo automated checks and will then be prepared for fabrication. If you want to manufacture the boards yourself, at home, then photoetching is the way to go.

Photoetching

Photoetching requires an ultraviolet (UV) light box, presensitized copper-clad board, developer, and etchant. This is quite practical for single-sided boards but requires more care for double-sided boards, where you need to align both sides accurately.

WARNING *Photoetching uses noxious chemicals as well as ultraviolet light, which, while not seeming bright, can do all sorts of damage to your eyes. Always observe the safety precautions specified on the equipment and chemicals that you use.*

Photoetching uses a transparency with an image of the PCB to be created printed onto transparency film that is then placed over copper-clad board that has been presensitized. These boards are not much more expensive than plain boards. The board is then exposed to UV light through the transparency film.

Figure 6-4 Homemade photoetching kit.

The board is then put into a tray of developer, and the image of the PCB tracks will become visible on the board just like an old-fashioned photograph being developed. Next, the board is etched in a chemical that dissolves the copper except where it is protected by the photographic image of the PCB tracks. Figure 6-4 shows the author's home-made setup for photoetching.

Rather than run the CAM processor, because there will be no solder mask, silk screen, or other refinements, you can set the layers to just display "Bottom" and then print the board, selecting the options for "Solid" and "Black." This is then printed onto transparency film (Figure 6-5).

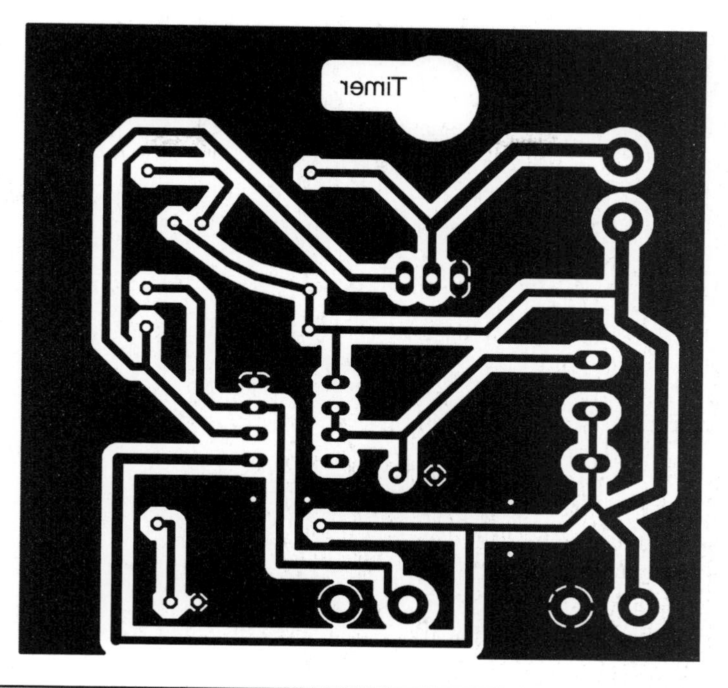

Figure 6-5 Printing the layout.

The protective film is then peeled off the copper-clad board, and I use a clip frame designed for photographs to press the transparency against the board while it is exposed in the UV light box.

Having been exposed, the board then needs to be put in developer, at which point the pattern on the board will start to appear. When development has finished, the board is placed in etchant (usually ferric chloride) that dissolves away the copper not protected by the developed image.

Your etchant will last longer the less copper is dissolved from the board, so use ground planes wherever possible.

When the board is finished, it will need to be drilled (if you are using a through-hole design), for which you will need a very fine drill bit. A diameter of 0.8 mm is ideal.

Milling PCBs

Low-cost desktop computer numerical control (CNC) routers offer a chemical-free method of producing PCBs by using a normal copper-clad PCB but then using a computer-controlled CNC router to cut away the unwanted copper (Figure 6-6).

FIGURE 6-6 CNC router cutting a PCB.

The process is similar to the photoetching method. Once the PCB artwork is done, the copper layer is dispatched to the router as if it were a printer. It suffers from the same disadvantage that double-sided boards are tricky. Because the copper has to be milled off the board where it is not required, this is another technique that benefits from a ground plane.

Toner Transfer

Another approach to homemade PCB manufacture is toner transfer. In this approach, the PCB layout is printed onto glossy paper in a laser printer. It is then ironed onto the copper-clad board using a clothes iron (turn off the steam setting).

The toner then provides sufficient protection to the board to allow it to be etched in the same way as photoetching.

Summary

I still get excited when a bubble-wrap package arrives with a set of shiny PCBs ready for me to use. Chapter 7 will look at the next step of soldering the conventional through-hole designs, hand soldering SMD PCBs, and cooking your PCBs in an oven.

CHAPTER 7

Soldering

In this chapter, I will first give a simple introduction to soldering of through-hole PCBs and then look at the more difficult but still perfectly possible soldering of SMD boards.

Tools

You do not have to spend a lot of money on tools for soldering PCBs. You can get perfectly good results with low-cost equipment. You wouldn't learn the violin on a Stradivarius, so don't buy a top-end Weller soldering station as your first soldering iron. Gradually improving your tools is one of the joys of electronic construction. Where would the fun be if you had the best of everything from day one?

General Tools

However you plan to do your construction, you will need certain tools when soldering a PCB.

Snips and Pliers

Snips (essential) are used to cut the excess leads off components after they have been soldered. They are sharp and allow you to get close to the solder joint. They are also useful for stripping the insulation off wire. Eventually, snips lose their sharpness and become blunt, especially if they are abused by being used to cut steel guitar strings. I use very cheap snips (Figure 7-1) and then replace them as soon as they become blunt enough to be irritating to use.

FIGURE 7-1 Snips and pliers.

Long-nosed pliers last longer and are generally a useful tool to have around. They can be used to grip a wire tightly while you strip the insulation from it or for holding onto a components that you are trying to desolder from a board without burning your fingers.

Multimeter

In a perfect world, everything would work the first time when you powered it up. The reality is that life is not quite like that. A multimeter (Figure 7-2) is an essential tool that will allow you to diagnose problems with your designs.

You do not need to spend a lot of money on a multimeter. A basic entry-level multimeter costing just a few dollars will do just fine most of the time. The most important setting that you will use most of the time is direct-current (dc) volts in a range of 0 to 20 V.

It is also useful to have a dc current setting of up to a few hundred milliamperes and a continuity test that buzzes when the test leads are connected together. Everything else is just bells and whistles that you might use once in a blue moon.

Most of the time, accuracy is pretty irrelevant too. When things go wrong, it is usually a matter of orders of magnitude. Thus, if your multimeter indicates a current of 10 mA when the current is actually 12 mA, that is usually good enough. It's when the current is 100 mA and you were expecting 10 mA there is a real problem.

Figure 7-2 Multimeter.

Soldering Station

Although you can get by with a soldering iron that plugs directly into an alternating-current (ac) outlet and has no way of adjusting the temperature, it is worth spending a few extra dollars on something that is thermostatically controlled and can accept fine-pointed tips (Figure 7-3). Make sure that you avoid anything advertised as being suitable for plumbing use.

When you are buying your soldering station, bear in mind that eventually the tips (also called *bits*) will need replacing. Make sure that replacements will remain available, or buy them when you buy the iron. With the trend that components are getting smaller and smaller, you probably will want a tip of perhaps 2 mm. There are many different tip shapes, and choice is a matter of personal preference. Many people prefer a chisel-shaped tip. A simple conical tip is another popular choice.

FIGURE 7-3 Low-cost soldering station.

If you plan to use lead-free solder, then temperature control is a must. You can get away with a simple low-cost soldering iron if you are using solder with lead in it because this type of solder is much easier to work with (see next section).

WARNING *It should go without saying that soldering irons get hot enough to burn your skin. Be very careful, and always put the soldering iron back into its holder as soon as you have finished with it. Do not leave it on the desk to roll off, triggering the automatic instinct to try to catch it when inevitably its lead gets snagged and it falls off the desk. Soldering also produces fumes from the rosin flux. It is a good idea to solder next to an open window or use a fume extractor.*

Solder

Traditionally, solder (Figure 7-4) has been made from tin and lead. Usually, this is in the proportion of 60 percent tin and 40 percent lead. The solder looks like a solid metal wire, but actually will normally have a core of flux rosin that helps the lead to flow when it melts. Legislation on the use of toxic chemicals has resulted in a reduction in the use of lead-based solder in favor of lead-free solder.

This type of solder is an alloy of tin, silver, copper, and small amounts of other metals. It looks like lead solder and still has a rosin core but is somewhat brittle

FIGURE 7-4 A reel of 0.7-mm tin, lead solder.

and has a melting temperature of about 200°C (392°F) versus about 190°C (374°F) for leaded solder.

The differences do not end there. Many people find lead-free solder much harder to work with. It does not flow as easily as lead solder.

Lead solder is still widely available, and unless you are producing a product that you are going to sell, it is really a matter of personal preference which type of solder you use. I know electronics enthusiasts who have a roll of lead-free solder that they use most of the time and then a roll of the "good stuff" (lead solder) that they use when they have something tricky to solder.

Whatever type of solder you use, you will have another choice to make: the gauge of the solder you buy. Two popular sizes are 0.7 and 1.2 mm in diameter. Use 0.7-mm or similar solder when integrated circuit (IC) leads are close together, for example, because it is much easier to use. If you need to solder some large terminals, you will find yourself feeding in quite a length of the narrow solder to deliver the required amount, but this is not really a problem.

Desoldering Braid

Desoldering braid (Figure 7-5) is not an essential tool for soldering, but it can come in very handy from time to time. As well as its primary use for "unsoldering" components, it is also great for mopping up excess solder, especially when hand soldering SMDs.

Figure 7-5 Desoldering braid.

The braid is made of copper impregnated with flux that encourages the solder to flow. Thus, when you place it between the pad from which you want to remove the solder and the soldering iron tip, it soaks up the solder like a sponge. Having done this, that section of braid cannot be reused, and you snip it off and throw it away.

Tip Cleaner

When you solder, it is very important that the tip of the iron is clean, or you will end up with blobs of solder that do not make a good joint. There are two methods of cleaning the tip, both used with the soldering iron hot. One is to use a damp sponge, and many soldering stations include a sponge holder. The other is to use a container of brass shavings, rather like a scouring pad (Figure 7-6).

The only real advantage of using a damp sponge is that it makes a great hissing noise as the hot tip of the iron is rubbed across it. It does, however, suffer from a number of disadvantages:

- The thermal shock of cooling the tip quickly as it comes into contact with the wet sponge will shorten the life of the tip.
- You have to keep wetting the sponge and need a supply of water.

Tools for Surface-Mount Devices

When attempting surface-mount soldering by hand, you probably will need all the equipment that I have just described for through-hole soldering. In fact, you can get away with just using regular soldering equipment. However, there are a number of special items that make surface-mount soldering simpler.

FIGURE 7-6 Brass soldering iron tip cleaner.

Hot-Air Gun

A hot-air gun (Figure 7-7) has interchangeable nozzles of different sizes that allow you to deliver a stream of hot air to an area of a circuit board.

You can normally set both the temperature of this air and the flow rate. You will need to control the flow rate because many SMD components are so small that they can easily be blown away by the pressure of air from a hot-air gun.

Solder Paste

When soldering SMDs, you can, with care, use regular solder. However, if you are using a hot-air gun or a reflow oven, then you will need to use solder paste (Figure 7-8). Solder paste is available in both lead-based and lead-free varieties and suffers the same pros and cons as those variants of regular solder.

Solder paste is made from microscopic spheres of solder in a suspension of flux. For industrial use, it is supplied in tubs, but for small-scale use, you can buy it in syringes ready for hand use. Solder paste should be kept in a refrigerator but warmed up to room temperature when you are read to use it.

WARNING *Solder paste is a liquid, and if you are using lead-based solder paste, it will easily find its way into the pores of your skin if you get it on your fingers.*

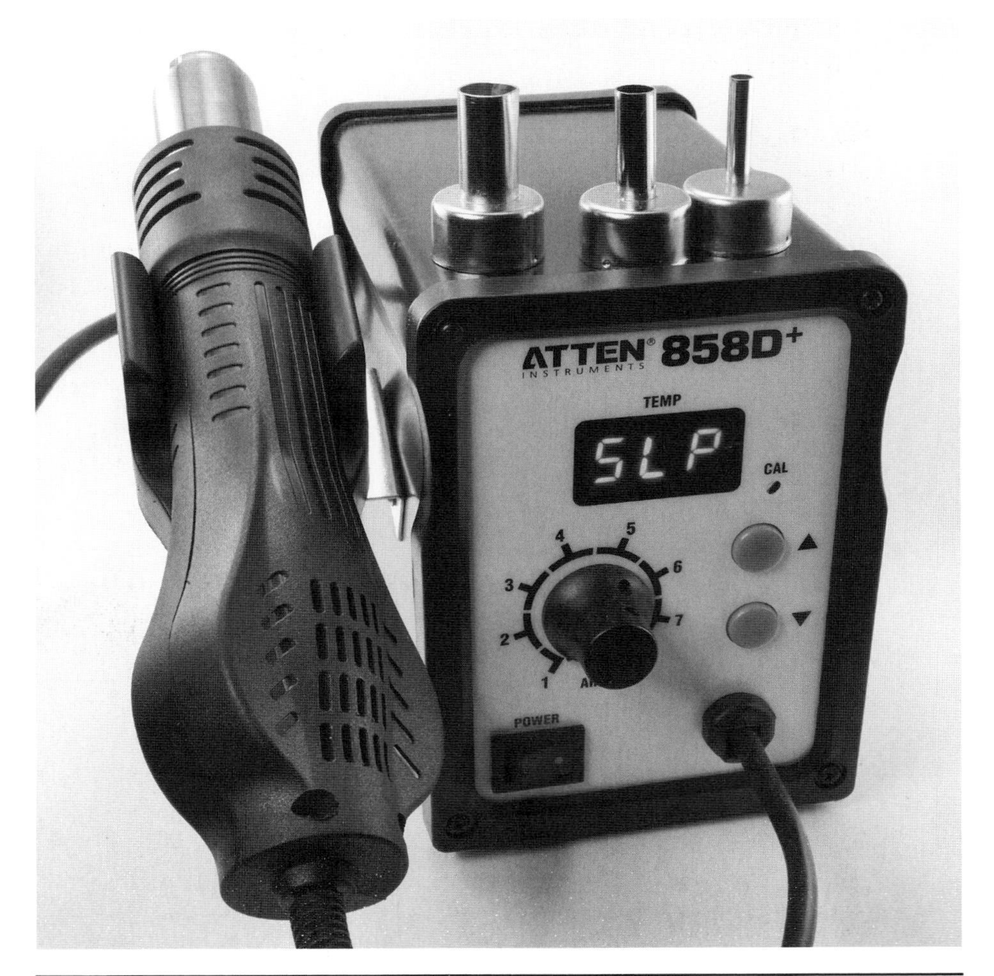

Figure 7-7 Hot-air gun.

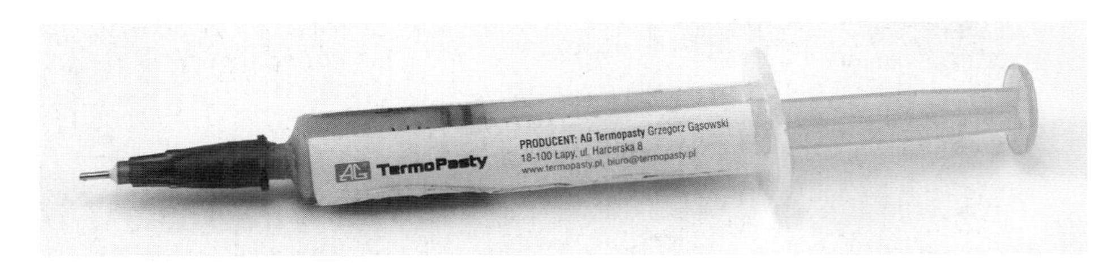

Figure 7-8 Syringe of solder paste.

Always wear latex gloves if, like me, you are a bit messy and likely to get it on your fingers.

Tweezers

To be able to pick up and place SMDs onto a board, you will need tweezers (Figure 7-9).

The tweezers should have a fine point and, most important, be nonmagnetic. If they are even slightly magnetic, then SMDs will stick to them because many contain ferrous metals.

Magnifier

It can be really hard to see what you are doing when you are working with surface-mount technology (SMT). A large magnifying work lamp such as that shown in Figure 7-10 can be a great help.

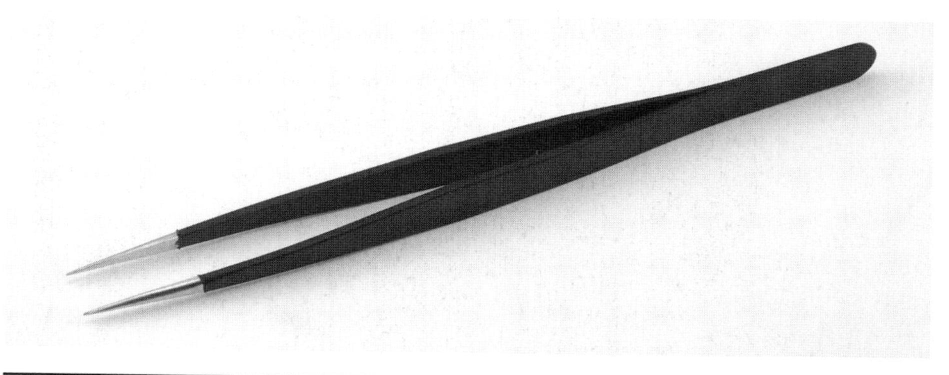

FIGURE 7-9 Nonmagnetic tweezers.

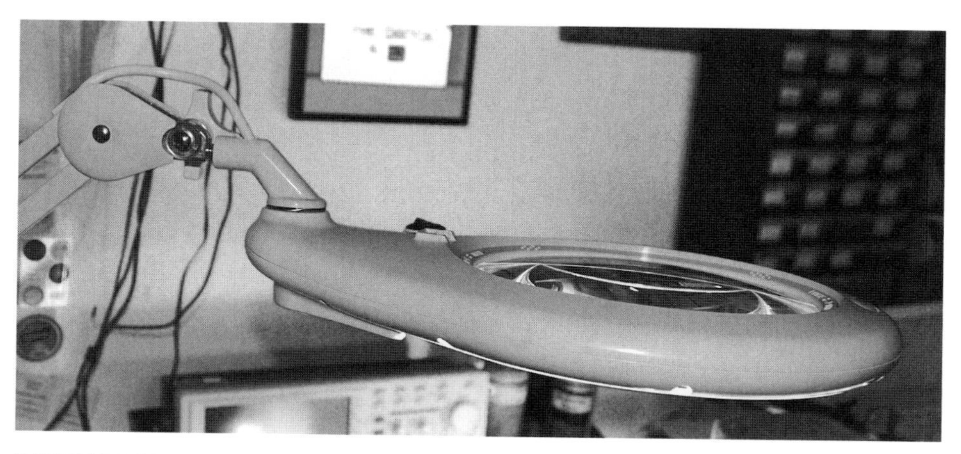

FIGURE 7-10 Magnifying work lamp.

These devices have a lighting ring around the lens that evenly illuminates the board on which you are working. Because you are looking through the lens with both eyes, all-important depth perception is preserved.

Some people take this a stage further and use a binocular microscope. These are available specifically for working on circuit boards, and a zoom version will allow you both to work on boards and to inspect them very closely for any problems. You should look for something that magnifies between 5 and 20 times.

Reflow Oven

When developing single boards, hand soldering works okay. It is a little tedious and time-consuming but can be done. The professional way to attach SMT components to a board is to use a reflow oven.

The basic idea is that you put solder paste on the pads of the board, place the components onto the pads, and then bake the entire board in an oven to melt the solder paste and attach the components. We will see how to do this in a later section.

Commercial reflow ovens are quite expensive, but many people make their own using low-cost toaster ovens, such as the modified device of the author's shown in Figure 7-11.

FIGURE 7-11 Modified toaster oven.

WARNING *These types of toaster ovens are often called "fire starters" for good reason. They are very simple designs, with little in the way of thermal insulation. This means that they get very hot, and if you modify them, they can become even more dangerous. If you decide to make one of these, never leave it unattended or anywhere near anything that could burn.*

The model shown in Figure 7-11 has been modified to replace the thermostat with a proportional power control module, and a digital thermometer has been added to allow the necessary accurate monitoring and control of temperature.

Soldering Through-Hole PCBs

Having explored the various tools we will need, let's start by learning how to solder through-hole PCBs. It is a good idea to try to follow these instructions on a PCB. You may wish to order one of the PCB designs from earlier in this book. We will use the sound meter design. Having a batch of PCBs means that you can sacrifice one or two on which to practice your soldering.

Through-Hole Soldering Step by Step

The first thing to do is to turn on your soldering iron and set the temperature. You will find conflicting advice for temperatures to use, but I set my soldering iron to 280°C (536°F) for lead-based solder and to 310°C (590°F) for lead-free solder. Once you get your eye in with soldering, you probably will want to work at a higher temperature, where the solder melts a bit more quickly. The higher temperature will not damage the components as long as you are quick.

When the soldering iron is up to temperature, clean it on the damp sponge or brass tip cleaner. Once cleaned, it should look bright and silvery.

The key to soldering is not to heat the solder but rather to use the soldering iron to heat the place where you want to solder and then feed solder onto that junction so that it melts and flows over the pad and the component lead. Figure 7-12 shows the steps involved in soldering the resistor leads.

First, push the component leads through their holes, and turn the PCB on its back (Figure 7-12*a*); then hold the tip of the soldering iron to the junction of the pad and the lead. Next, feed solder into the joint so that it flows all around the lead and covers the pad (Figure 7-12*b*). Once it has flowed all around, you can stop adding solder and move the tip away. You do not want the pad to be heaped high with solder. The solder ideally should form a nice "mountain" of solder around the lead (Figure 7-12*c*). Finally, you can snip off the excess lead (Figure 7-12*d* and *e*).

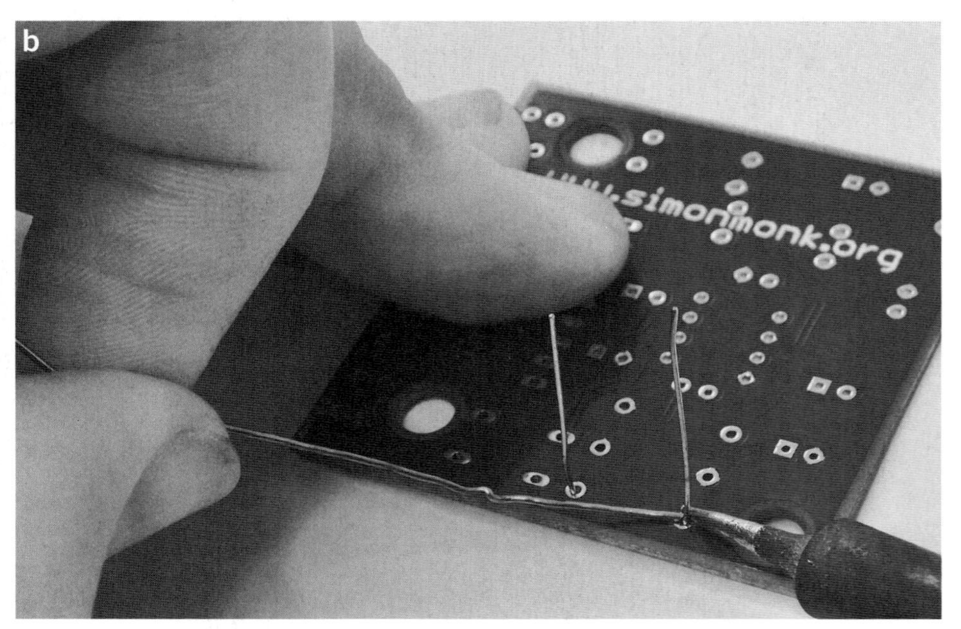

Figure 7-12 Soldering a resistor.

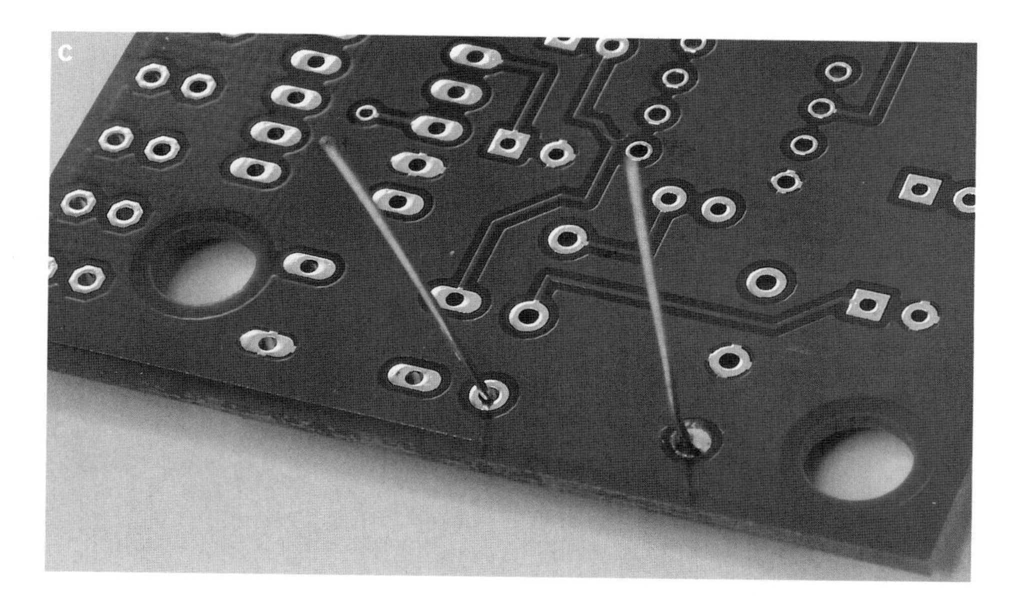

Figure 7-12 Soldering a resistor (*continued*).

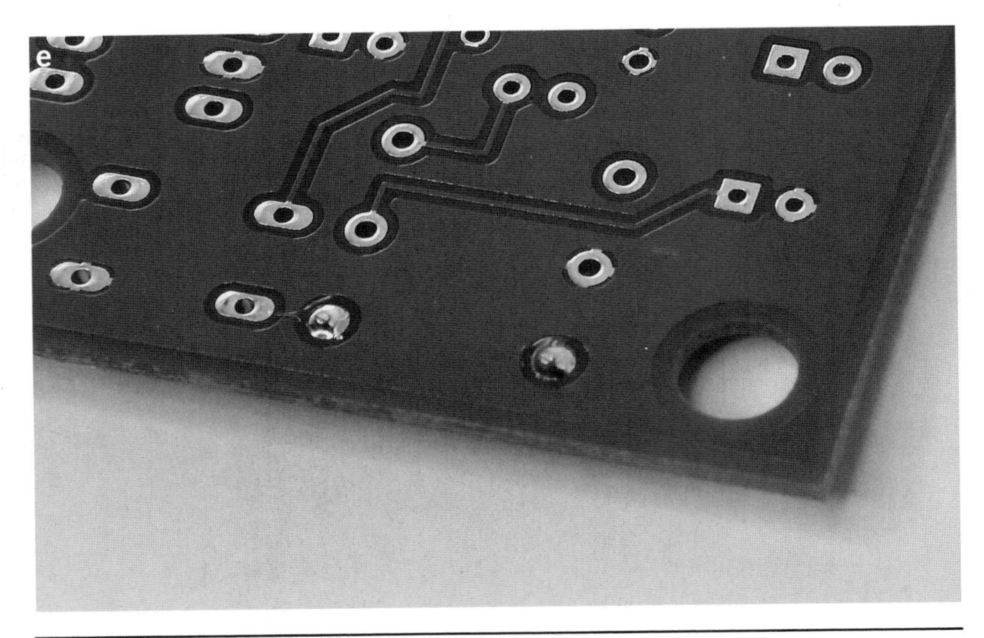

FIGURE 7-12 Soldering a resistor (*continued*).

When soldering components onto a PCB, you can make life much easier for yourself if you start with the components that lie closest to the surface of the board. In this way, when you turn the board on its back, the weight of the board will keep the components pressed against the board.

If you make a mistake and find yourself needing to desolder a joint, then Figure 7-13 shows the steps you should take.

If the board was soldered some time ago, then you might find that it is quite hard to desolder. Actually heating it and adding a bit of solder could make it easier to desolder the component. Thus, after optionally resoldering the joint as described earlier, place an unused end of the soldering braid against the joint (Figure 7-13*a*), and then press it down onto the solder pad using the tip of the soldering iron (Figure 7-13*b*). As the solder melts, it should be drawn into the solder braid. You will be lucky to draw off all the solder in one go, so most likely you will need to snip off the now-solder-covered end of the braid and repeat the process of pressing it against the joint. Eventually, you should have most of the solder removed, and the joint will look like Figure 7-13*c*.

Repeat this for the other lead or leads of the component. If you are very lucky, you will just be able to wait until the component has cooled and then gently pull the component back through the hole from the top. However, it is more likely than not that there will still be a little solder holding the component in place. If

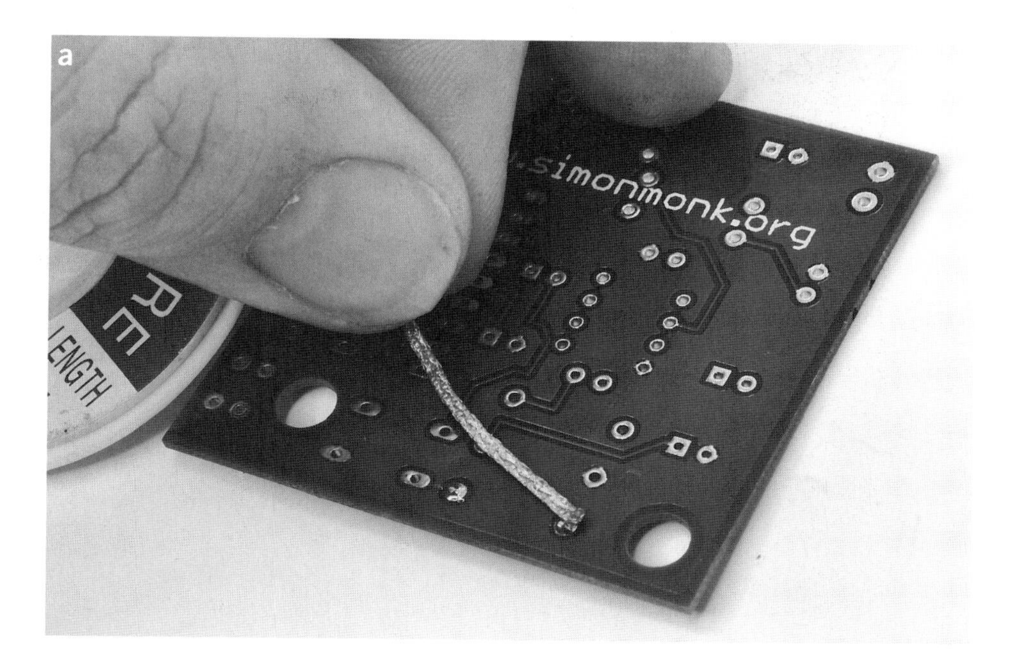

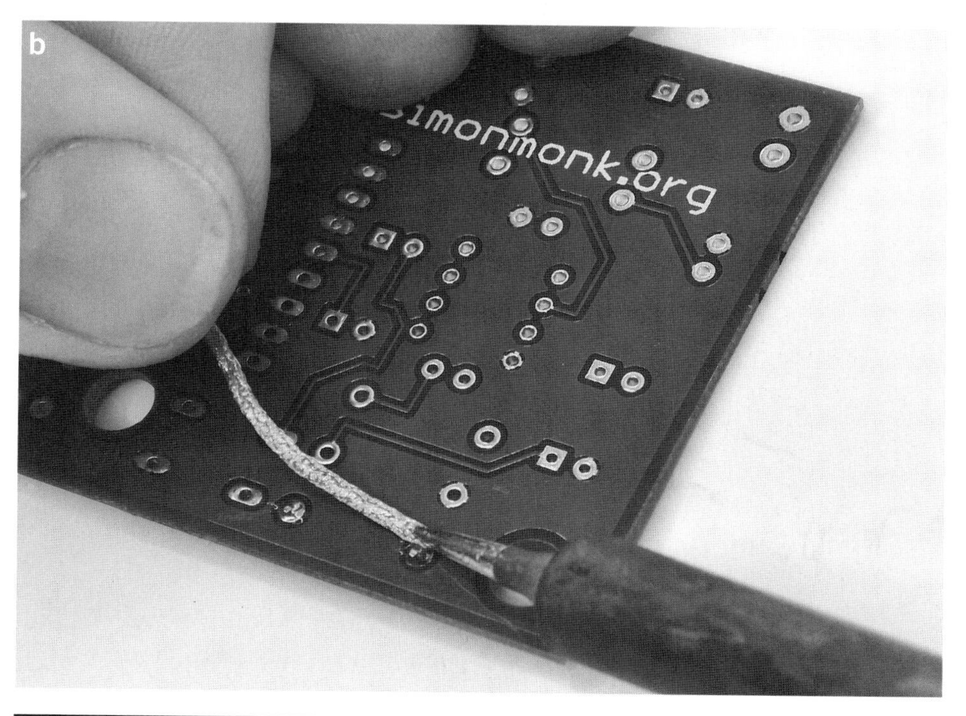

FIGURE 7-13 Desoldering.

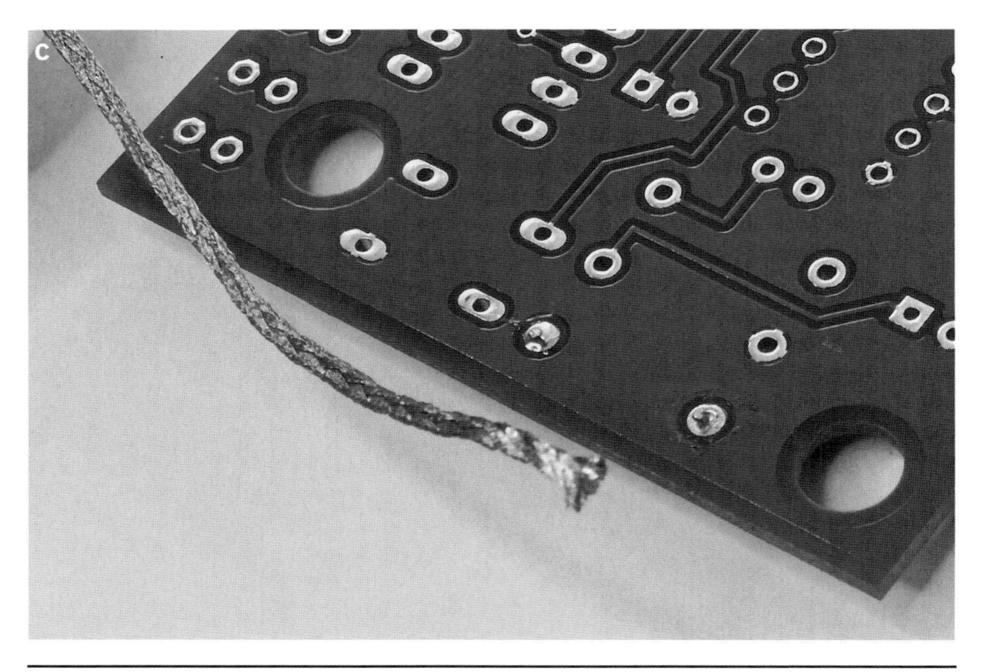

FIGURE 7-13 Desoldering (*continued*).

this is the case, then hold one component lead from the top side with long-nosed pliers, and heat the pad from underneath while applying a gentle pull on the lead to pull it back through. If you cannot get to the lead and you don't mind breaking the component, then snip it in half or snip the lead off on the top to make it easier to pull through.

All this means that you are likely to be heating the board for a while, which can make it look scruffy or even damage the board. Pulling the lead through with force is also likely to damage the PCB, and ultimately, too much heat will eventually cause the pad to separate from the board.

SMD Hand Soldering

As long as the SMDs you use are at the larger end of the size scale and have pins on the edges of the device rather than on the underside, then they can be soldered relatively easily using a regular soldering iron. The main problem is that they are so light, and because they do not have leads projecting through the holes, there is nothing to hold them in place when you try to solder them.

Soldering Two- and Three-Legged Components

The sequence of steps to solder a 1206 resistor into place is shown in Figure 7-14.

This is one of those situations where you would really benefit from having three hands—one to hold the soldering iron, one to hold the tweezers to keep the SMD in place, and one to apply the solder. If you do not have three hands, then a good trick is to place a small mound of solder on one of the pads (Figure 7-14*a*) and then, while holding the SMD in place with the tip of your tweezers, press its lead into the little solder mound (Figure 7-14*b*) with the tip of your iron. The SMD will now stay in place without the need of the tweezers as you solder the other end normally (Figure 7-14*c*). It's then usually a good idea to touch the first end with the iron and a little solder just to freshen it up.

An alternative technique is to place solder paste onto the pads, hold the SMD in place with the tip of your tweezers, and then touch the tip of the iron to each lead until the solder paste beneath melts.

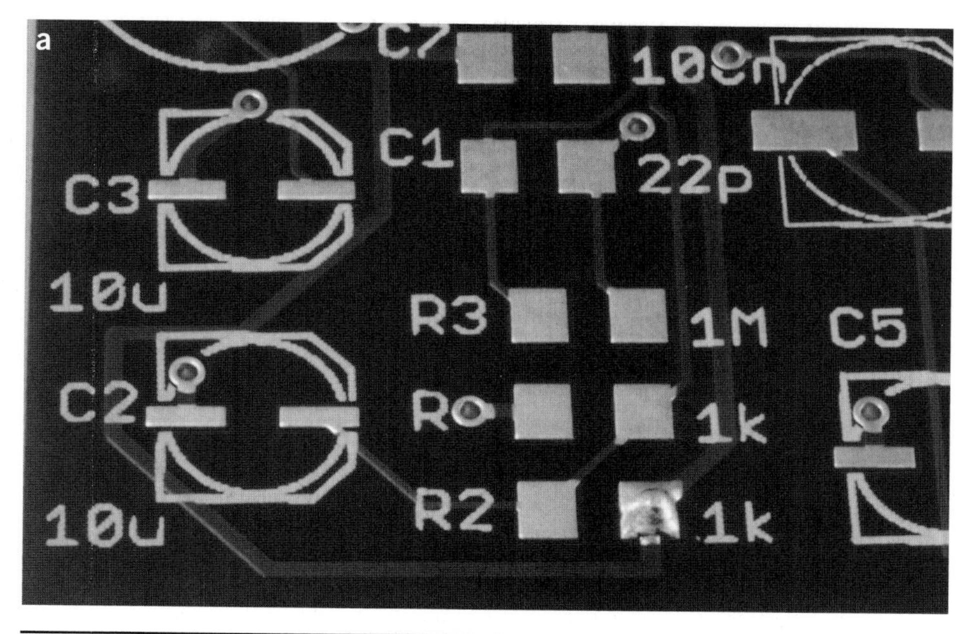

FIGURE 7-14 Using a soldering iron on an SMT resistor.

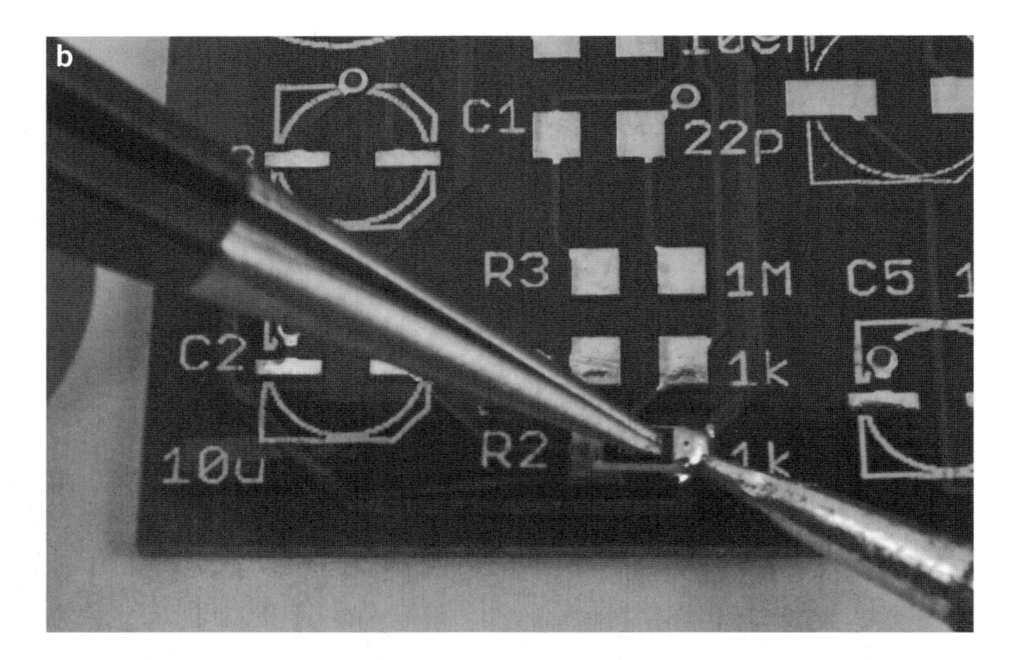

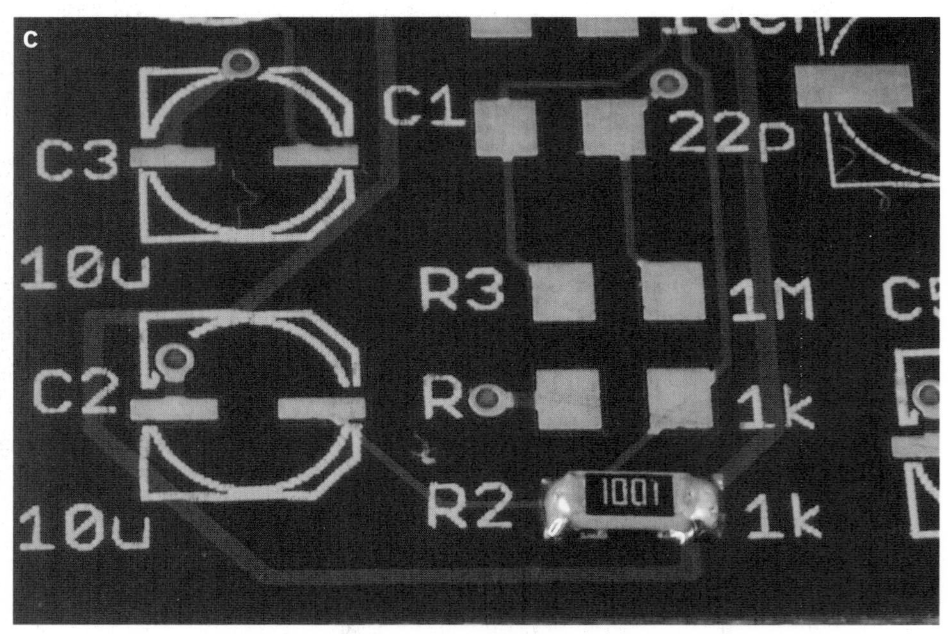

FIGURE 7-14 Using a soldering iron on an SMT resistor (*continued*).

Soldering IC Packages

The preceding approach works just fine for two- and three-legged devices, but when it comes to ICs, the process can become more tricky. You can try the conventional soldering iron approach, pinning the IC down with one corner pin and then carefully soldering the remainder of the leads, but often the solder will make unwanted bridges between the pins.

A good trick is not to worry too much about those bridges, but when you have finished soldering, lay desoldering braid along the row of pins and heat along the whole length to remove the excess solder. This is shown in Figure 7-15.

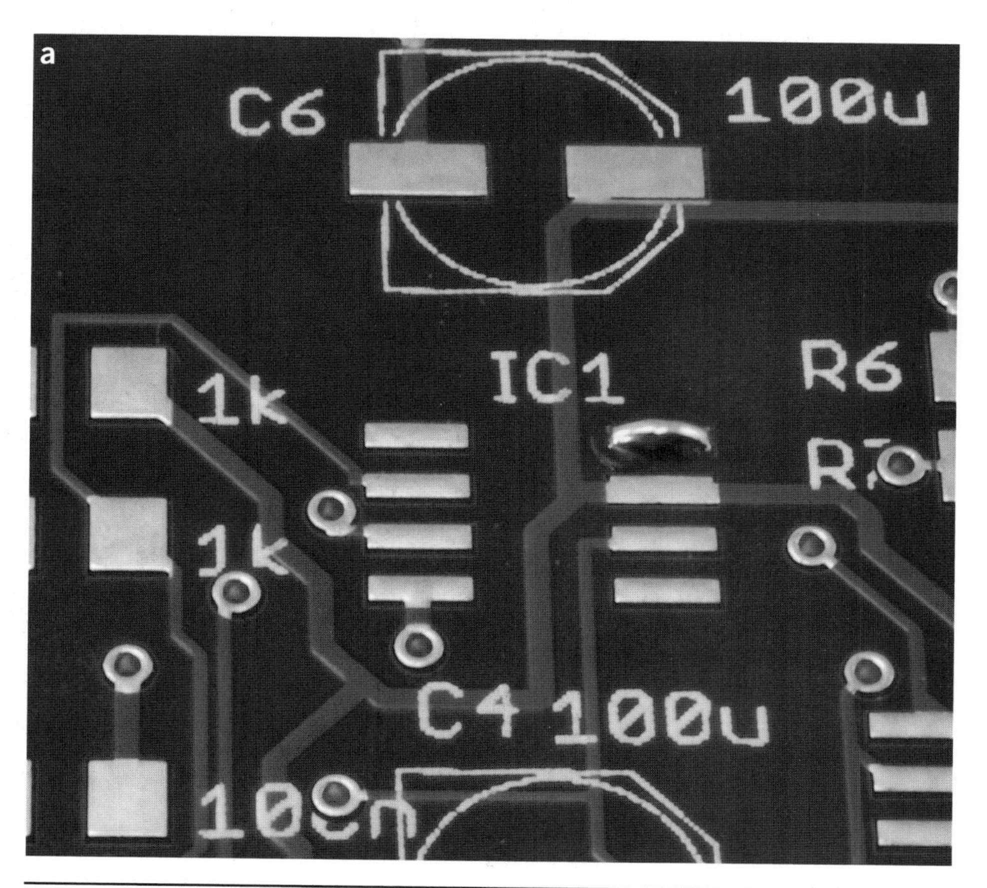

FIGURE 7-15 Hand soldering an SMT IC.

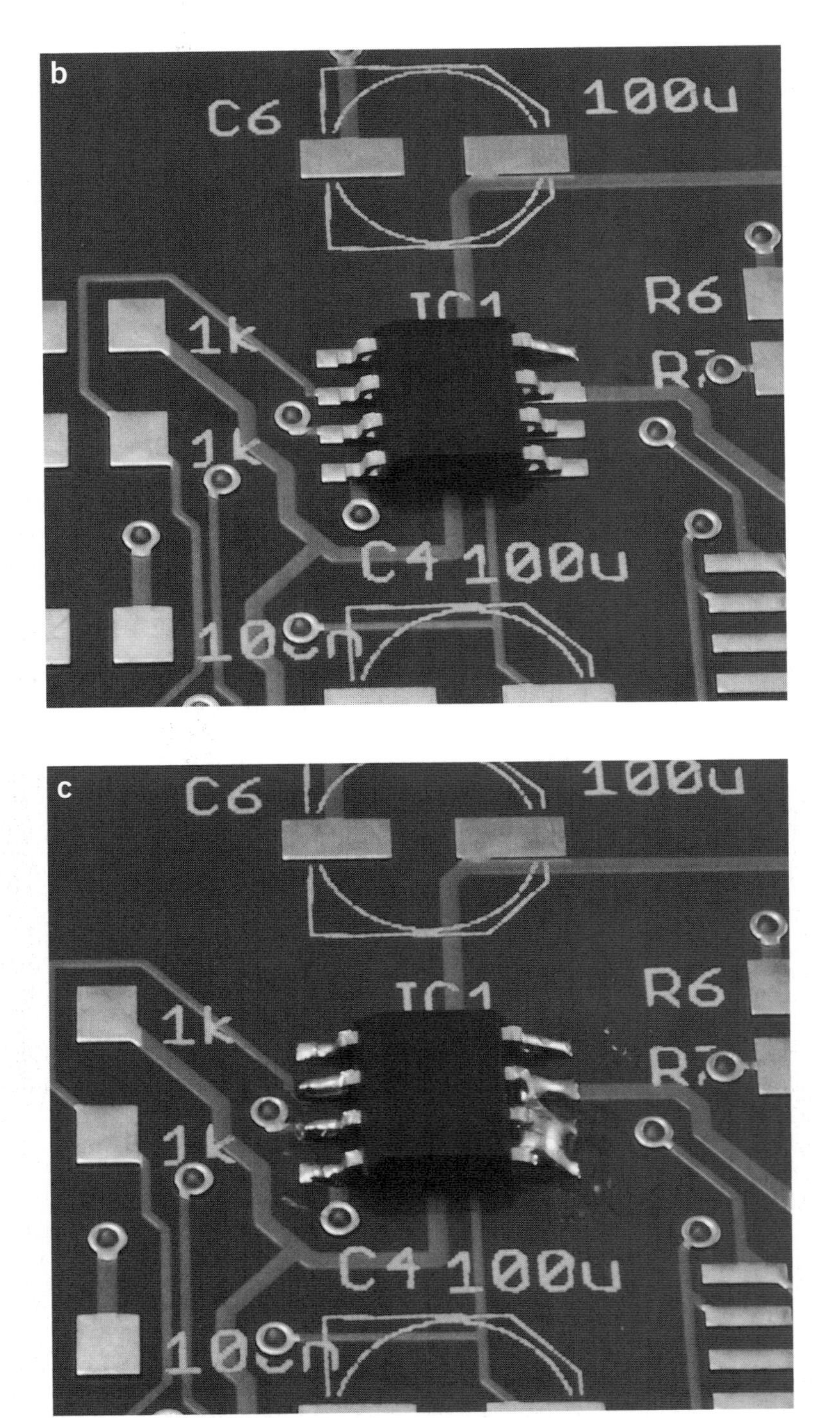

FIGURE 7-15 Hand soldering an SMT IC (*continued*).

SMT with Hot-Air Gun

Generally speaking, it is much easier to solder SMDs with solder paste and a hot-air gun than with a soldering iron.

Soldering Two- and Three-Legged Components

The steps for soldering with solder paste and a hot-air gun are illustrated in Figure 7-16.

First, place a small amount of solder paste on each pad. You can either squeeze it out through the syringe needle or use clean gooey paste, wiping away any crusty paste from the end of the syringe before you start (Figure 7-16*a*). Occasionally, I find a wooden toothpick to be useful for spreading the paste around a bit. Do not worry if it is a little untidy; when it melts, surface tension will cause it to pull back onto the solder pad.

Next, using your tweezers, place the component onto the pads (Figure 7-16*b*). Now fit a small nozzle onto your hot-air gun, set the temperature to 280°C

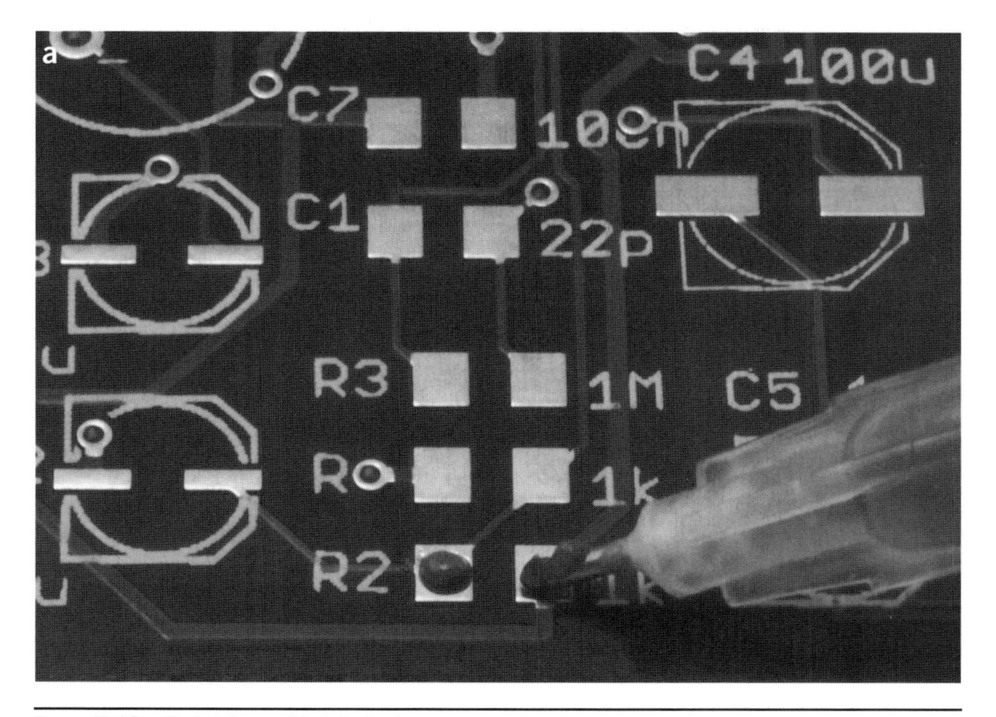

FIGURE 7-16 Soldering with a hot-air gun.

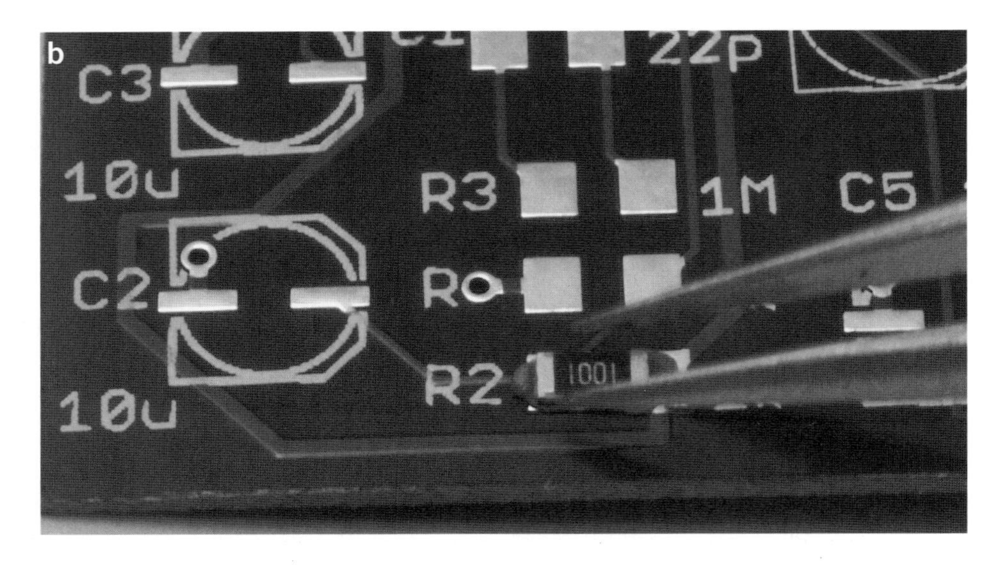

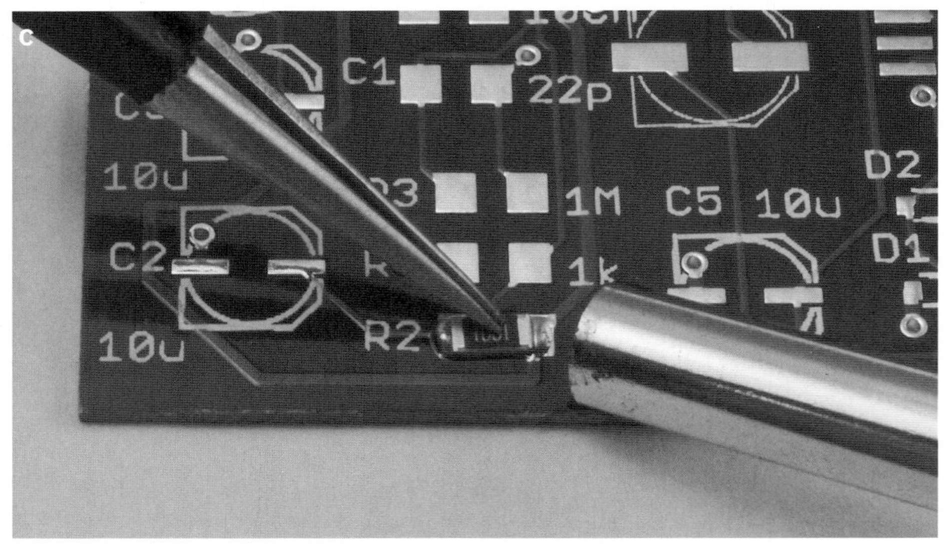

Figure 7-16 Soldering with a hot-air gun (*continued*).

(536°F) for lead-based solder and 310°C (590°F) for lead-free solder, and set the flow rate to perhaps one-quarter of full power. When the air gun is up to temperature, pin the component down with the tip of your tweezers, and then play the hot-air gun over the component and its leads until the solder paste melts (Figure 7-16*c*). When the solder melts, you will see it change from dull gray to shiny silver and see it spread across the pad. While still pinning down the component, put the hot-air gun safely back on its stand, and after a second or two, when you are sure the solder has set, let go with the tweezers.

If you do not hold the component in place with tweezers, then even at very low airflows, the hot-air gun probably will blow the component out of position.

Soldering IC Packages

Soldering ICs is very similar to the process just described. The only real difference is that you may well want to clean up the connections using soldering braid, as shown in Figure 7-15.

Packages with Hidden Connections

Some IC packages have inaccessible components on their underside. To solder these, put paste on the pads as normal, and then while you hold the IC in place with the tip of your tweezers, play the hot-air gun over the whole IC socket until you feel like the solder has melted. If you do this for too long, you may damage the chip.

Using a Reflow Oven

By far the quickest way to solder a SMT board of any complexity is to use a reflow oven. It has the advantage that you only have to place the components on top of the solder-pasted pads. Once this is done, the whole board is cooked in the reflow oven, soldering all the components in one go. What is more, once you gain confidence, you can cook a whole batch of PCBs in one go. In any case, the cooking process only takes a couple of minutes.

If you have a board that contains both surface-mount and through-hole components, then solder the surface-mount components first.

Get Everything Together

Solder paste will dry out after maybe half an hour, so before you do anything, make sure that you have all the components that you need and that you know exactly how they will fit onto the PCB (which way around LEDs and so on are to be placed). I find it useful to actually put the components onto the board without any paste just to make sure that I have everything I need. I have the board on a plain sheet of paper, so when I am sure that I have everything in place, I can move everything off to one side, keeping the same relative positions of the components.

Another approach is to use labeled bottle caps or even just circles on a sheet of paper labeled with the component part and/or value.

Applying Solder Paste

The low-tech way of applying solder paste is the same as I described earlier when I looked at using a hot-air gun. Simply go round the board adding a little blob of solder paste onto every pad on the PCB using the syringe dispenser. This is the time-consuming bit.

The alternative to using a syringe is to use a stencil. Many PCB manufacturing services will also (for a small extra fee) supply you with a stencil. This can be made of thin steel, Mylar, or other materials and is placed over the PCB. It masks out most of the PCB surface except for the areas where solder paste needs to be deposited. You then place some solder paste on the stencil and "squeegee" the solder paste into all the holes in the mask. The excess solder paste is then scraped up and the template removed, leaving solder paste on all the pads.

You can also make your own stencils, and if you search the Internet, you will find various do-it-yourself (DIY) techniques for doing this using laser or vinyl cutters or even transferring toner onto a cut-up drink can that is then put in acid to dissolve away the holes and make the template.

In this example, we are applying the solder paste by hand to the fairly densely packed PCB that has a wide variety of different component types. When the solder paste has been applied, the board will look something like Figure 7-17.

Note what a poor job the author has done in supplying the solder paste evenly. You should aim to be neater than this, but even with this level of messiness, it is still likely to work.

Populating the Board

Starting with the smallest, lowest-lying components and at the far end of the board, place components on the pads using tweezers until all the components are in place (Figure 7-18).

The board is now ready for cooking.

Baking the Boards

If you have a proper reflow oven, then baking the boards is pretty much as simple as putting them in and pressing a button. Behind the scenes, there is some fairly careful temperature control going on.

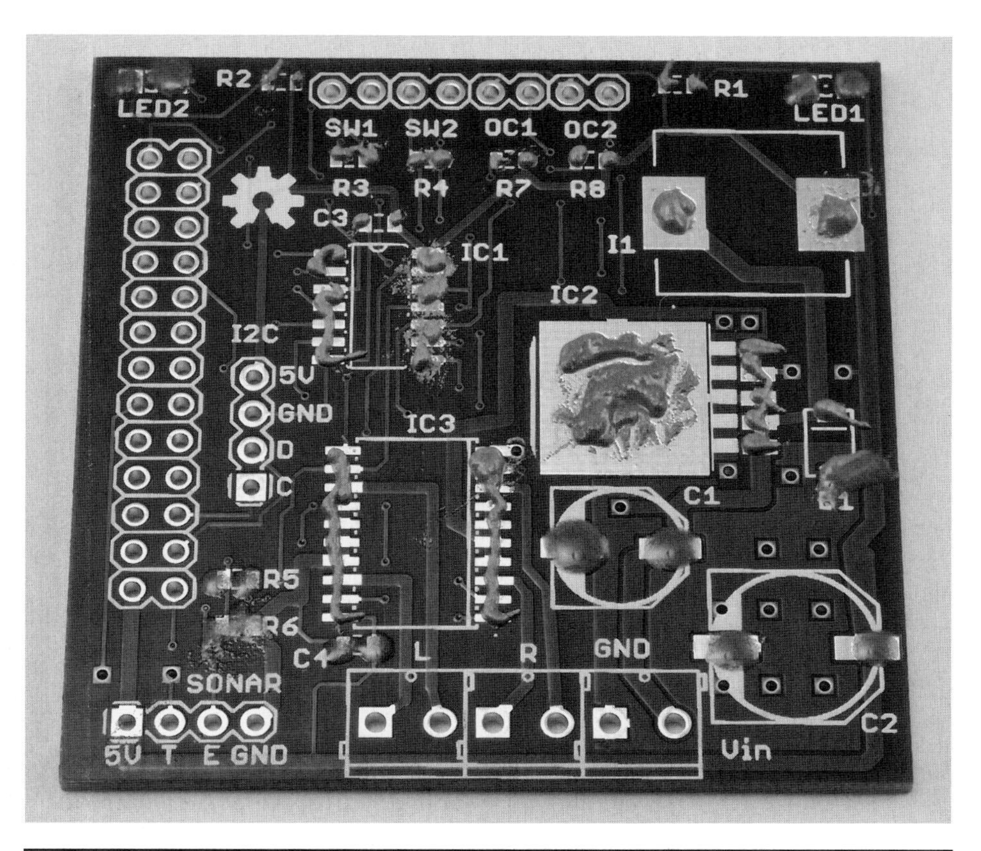

Figure 7-17 Board prepared with solder paste.

Solder paste requires a certain profile of changing temperature over time for it to do a good job of soldering components in a reflow oven. It has to go through four distinct stages:

- **Preheat.** Activate the flux.
- **Soak.** Warm the whole board to just below the solder melting point.
- **Spike.** Fast as you can, heat the board to above the melting point to reflow the board.
- **Cool.** Cool everything down before the components and board are damaged.

Each of this stages has precise temperatures and timings associated with it. A commercial reflow oven will allow you to select from preset profiles to match the paste you are using and control all aspects of the heating. Figure 7-19 shows the

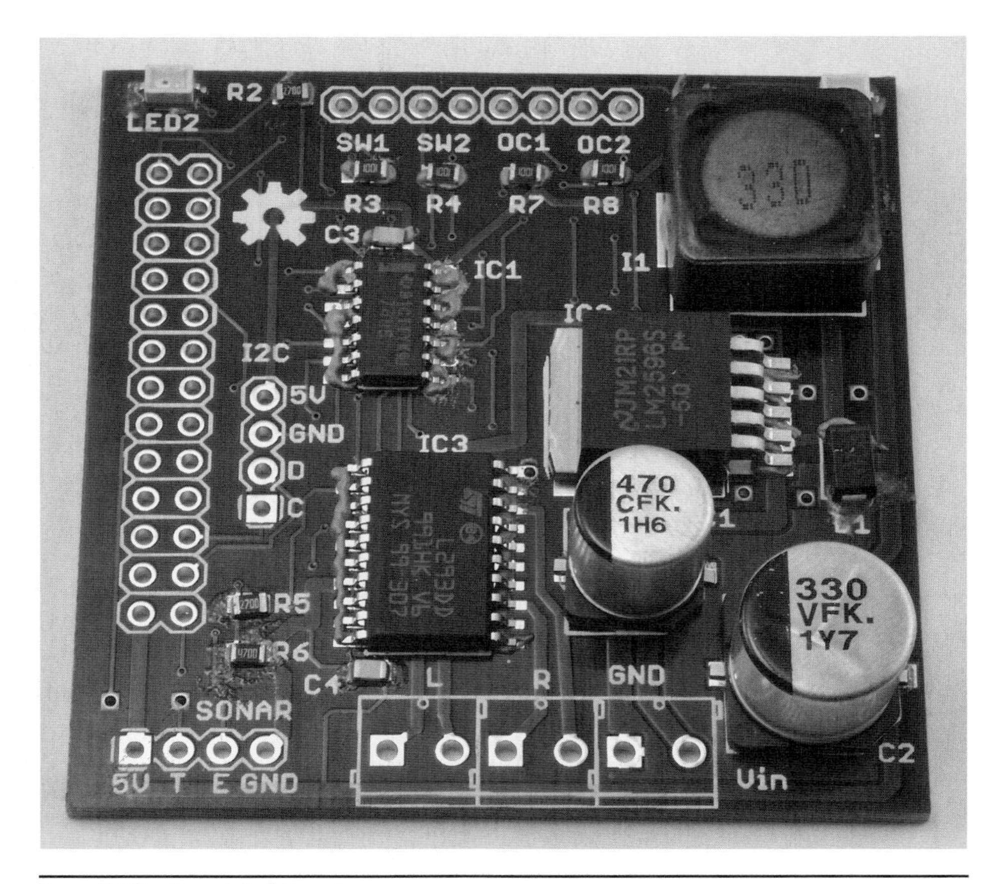

FIGURE 7-18 Populated board.

temperature profile of some leaded solder paste, and Figure 7-20 shows the board in the author's home-made oven.

Using a home-made device such as this will never be as reliable as using a professional oven, but it is fine for prototyping. Making one of these is dangerous and should only be undertaken if you really know what you are doing. You can find instructions for doing this at the following web pages:

- www.sparkfun.com/tutorials/60
- www.freetronics.com/pages/surface-mount-soldering-with-a-toaster-oven#.Us_cUGRdVyF

Both these tutorials describe how to manage the temperature by hand without the need for a complex controller. The final board is shown in Figure 7-21. You

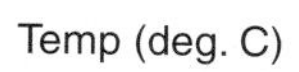

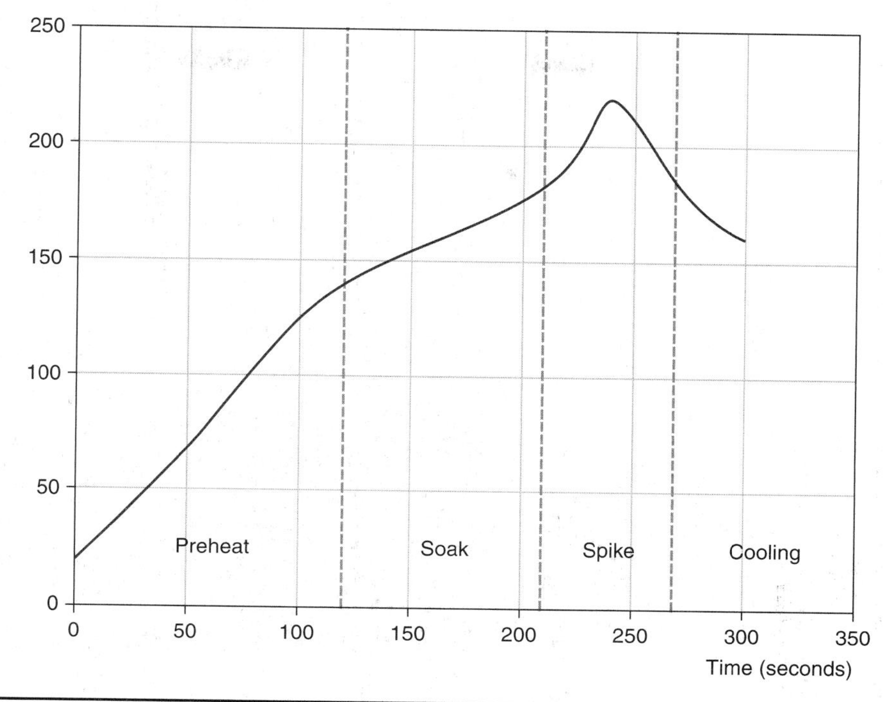

Figure 7-19 Leaded solder paste reflow profile.

Figure 7-20 Cooking the board.

FIGURE 7-21 Final board.

can see that the LED has moved a little. This is an effect of my cavalier application of solder paste but can be easily corrected with a little hand soldering.

Before you power up any board that you have made, you need to go over the whole thing very carefully with a magnifying glass to check that there are no accidental solder bridges and that all the pins are soldered to pads. Be especially careful around SMT ICs. You can mop up excess solder causing bridges using desoldering braid.

Depending on the solder paste that you used, you also may find little patches of flux and even tiny balls of solder on the board. I have a soft toothbrush that I use just to brush over the board. This will also highlight any loose components by brushing them off the board. This can be improved if you look for no-clean solder paste.

Summary

In this chapter we examined a few techniques for soldering components onto your PCB. Soldering is one of those skills that improves with practice, so never commit to too much on your first attempts. Start with something simple that you are prepared to throw away, and then work up. It is also not a bad idea to start with a simple soldering kit to get some practice.

In Chapter 8, we will follow an end-to-end example, designing and then building an Arduino shield.

CHAPTER 8

Example: An Arduino Shield

In this chapter, you will learn how to use EAGLE to design a plug-in shield for the popular Arduino microcontroller board. The design will use a mixture of SMDs and through-hole devices. By following this example, you will learn how to design your own Arduino shields.

The example developed is for a four-digit, seven-segment LED display with a few extras in the form of a real-time clock chip, a piezo buzzer, and a rotary encoder. The shield could be used by those experimenting with making their own alarm clock or timer of some kind.

Introducing Arduino

Figure 8-1 shows an Arduino Uno board. This is the most popular of the Arduino range of microcontroller boards.

The board has two rows of sockets on either side, and a wide range of plug-in shields are available that are the same size as the Arduino board but fit over the top of it and add extra features. Among other things, there are shields for Ethernet interfacing, controlling motors, and various kinds of displays.

Shield Design

When designing a shield, before wading into the schematic design, it is usually a good idea to think about exactly what the shield will do. In this case, we could set down the design goals of the shield as follows:

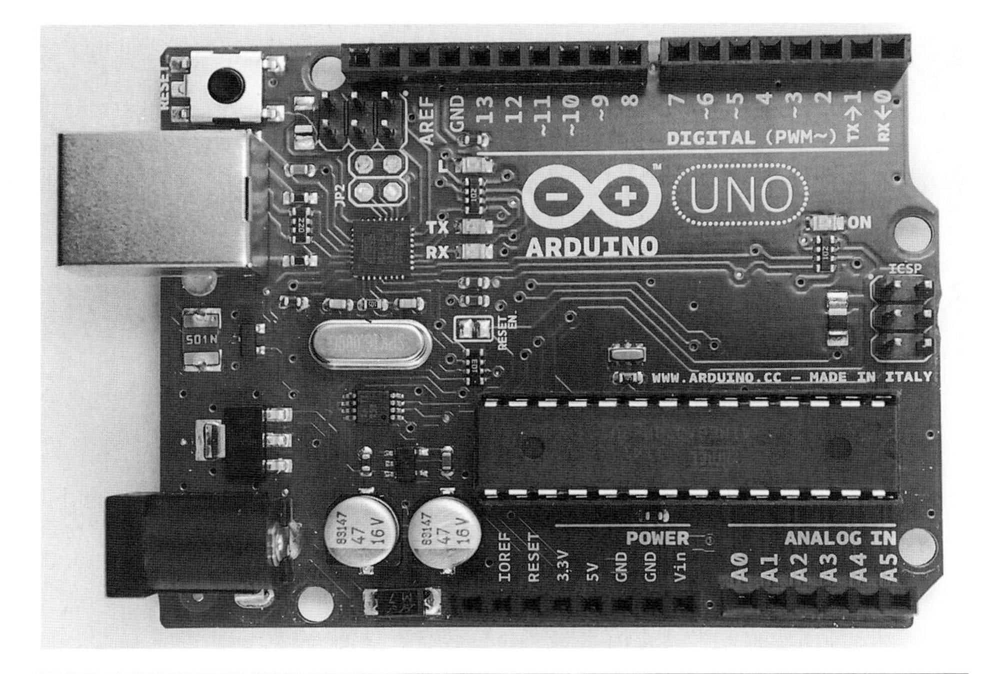

Figure 8-1 An Arduino Uno.

- A top shield; that is, no shields stacked on top of it
- Attractive design with retro four-digit, seven-segment display
- Real-time clock (RTC) chip
- Buzzer
- Rotary encoder with push switch
- Two-pin header output for connection to a relay module

Arduino R3 Shield Template

Adafruit has created an EAGLE part for making a shield for the latest version of the Arduino Uno (the R3 at the time of writing). To use this, you will need to download and install the Adafruit EAGLE library. This is well worth doing anyway because there are lots of other useful things in the library.

If you did not install this library back in Chapter 1, then download the file `adafruit.lbr` from https://github.com/adafruit/Adafruit-Eagle-Library/blob/master/adafruit.lbr. Right click on the "Raw" button, and save the file to the `.lbr` folder in your EAGLE installation folder. You will then need to restart EAGLE.

A Four-Digit LED Example

Now that we have clarified what the board will do, create a new project (called *TimerShield*) and then a new schematic within the project also called *TimerShield* (see Chapter 2).

Schematic

First, let's add a letter-sized frame into which we can add our components. This step is obviously optional, but it does lend a certain air of professionalism to the design and is expected if you are going to be releasing the design to others, say, as an open-source hardware (OSH) design, as is the case for this shield.

You can find the frame part in the Sparkfun Aesthetics library, where it is called `FRAME_LETTER`. Use the Text tool to add some text over the "Design by" and "REV" fields. Before you drop the text over the field, change the layer in the Action toolbar to be layer "94 – Symbols." You probably also want to set the font size using the "Size" dropdown in the Action toolbar.

My prefered way of designing a schematic is to first add all the key large components that I am going to need and then move and rotate them into well-spaced positions. I then add in all the smaller components such as resistors and capacitors.

In this case, this means adding the components listed in Table 8-1 to the schematic.

Table 8-1 Key Components

Part	Description	Search for
Arduino	Arduino R3	*ARDUINOR3-DIMENSION*
LED1	Four-digit, seven-segment LED display	*7-SEGMENT-4DIGIT-YOUNGSUN*; then select the "Long Pads" option (easier to solder)
IC1	DS1307 RTC IC	DS1307SO8
BAT1	3-V lithium battery holder	*BATTERY12MM*
SW1	Rotary encoder with switch	*ROTARTY_ENCODER*; it is misspelled in the library

After adding the components, set their names as shown in Table 8-1, and move them around and rotate them so that there is room for the extra components. Also add in the GND and 5-V power supplies. There are many of these in the libraries; I picked the ones from the Sparkfun Aesthetics library.

Aim for an arrangement that looks something like Figure 8-2.

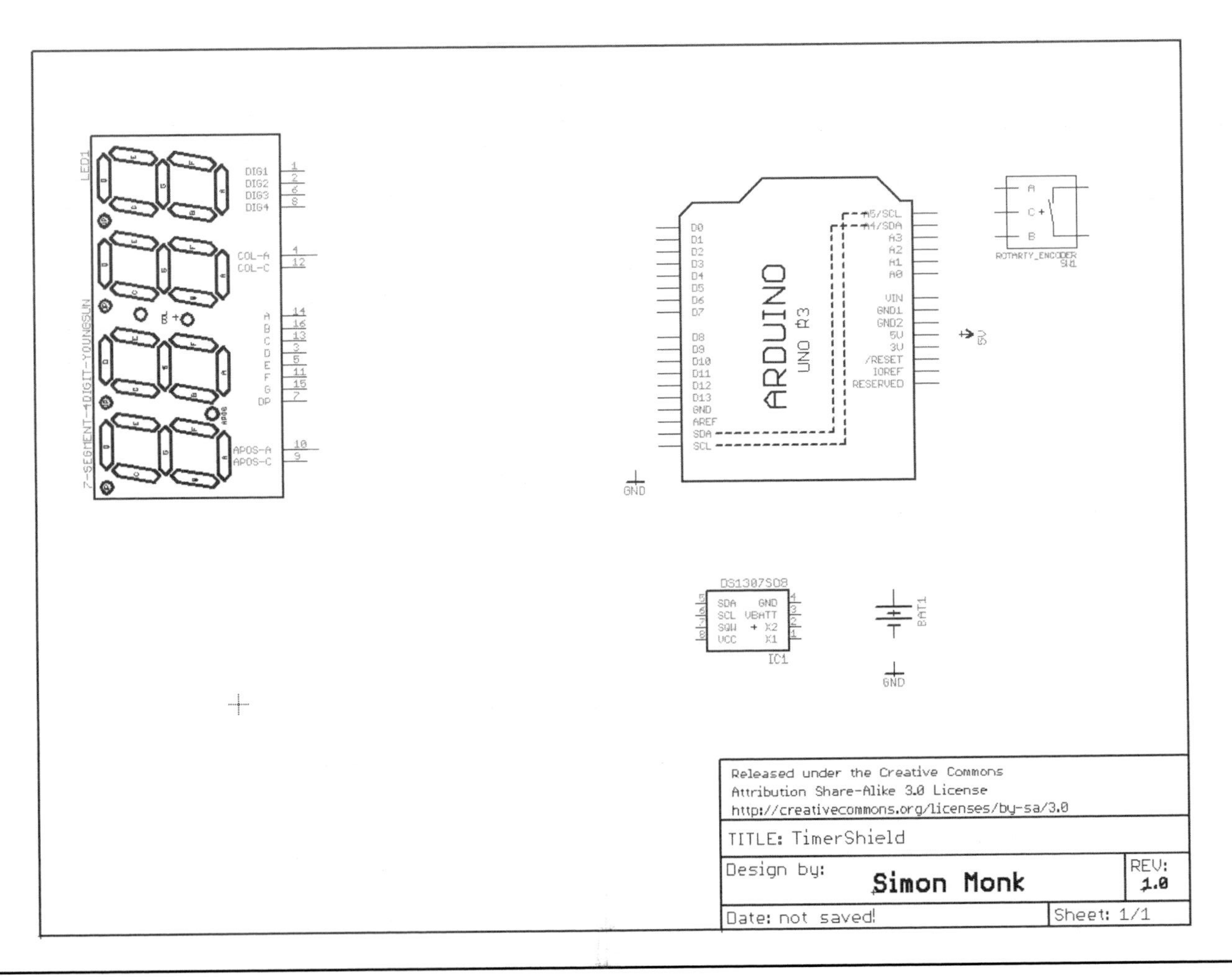

FIGURE 8-2 Initial schematic with key components.

Turning our attention to the more minor components, there is going to be an *NPN* transistor for each of the cathodes of the four digits and accompanying base resistors. There will also be current-limitting resistors for each of the eight anodes, as well as the apostrophy and colon anodes.

Add the components as listed in Table 8-2.

Table 8-2 Remaining Components

Part Name	Value	Description	Search for
Q1–Q4	—	MMBT2222A NPN SOT23 transistor	*MMBT2222*
R1–R4	1 kΩ	1-kΩ 0805 resistors	*RESISTOR0805*
R5–R14	270 Ω	270-Ω 0805 resistors	*RESISTOR0805*
R15, R16	4 kΩ	4.7-kΩ 0805 resistors	*RESISTOR0805*
X1	—	32.768-kHz crystal	*TC26H* and then the Adafruit version
SG1	—	Piezo buzzer	*BUZZERNS*
J1	—	Two-way screw terminal (3.5 mm)	*M023.5MM*

You will need to flip the transistors using the Mirror command so that they are facing the correct way around for the base to be closer to the Arduino. I also used the Smash command on the transistors to put the Name label closer to the symbol. Rather than add the four transistors using the Add command, it is quicker to add one transistor, mirror it, smash and move the label closer, and then use the Copy command to duplicate it three times. The same is true of the resistors of each value. So just add one resistor, set its value to 1 kΩ, and then copy it three times for R1 to R4, and then do the same for the other resistor values. Also change the values of the resistors as per Table 8-2.

When all the components are on the schematic, the design should look something like Figure 8-3.

We now just need to add all the nets to the schematic. For a schematic such as this, it is probably best to connect up the components in a certain area (perhaps around IC1) and until it looks like everything is connected that should be connected.

When working on a particular area, it helps to zoom in tight so that you can see what you are doing. Note that the display used is in real life slightly different from the library part. In fact, the colon and apostrophe LEDs have their anodes and cathodes swapped over. Thus, use Figure 8-4 as a reference. In Chapter 11, we will use this as an example of how you can modify a part.

Once all the nets have been wired up, the schematic should look like Figure 8-4.

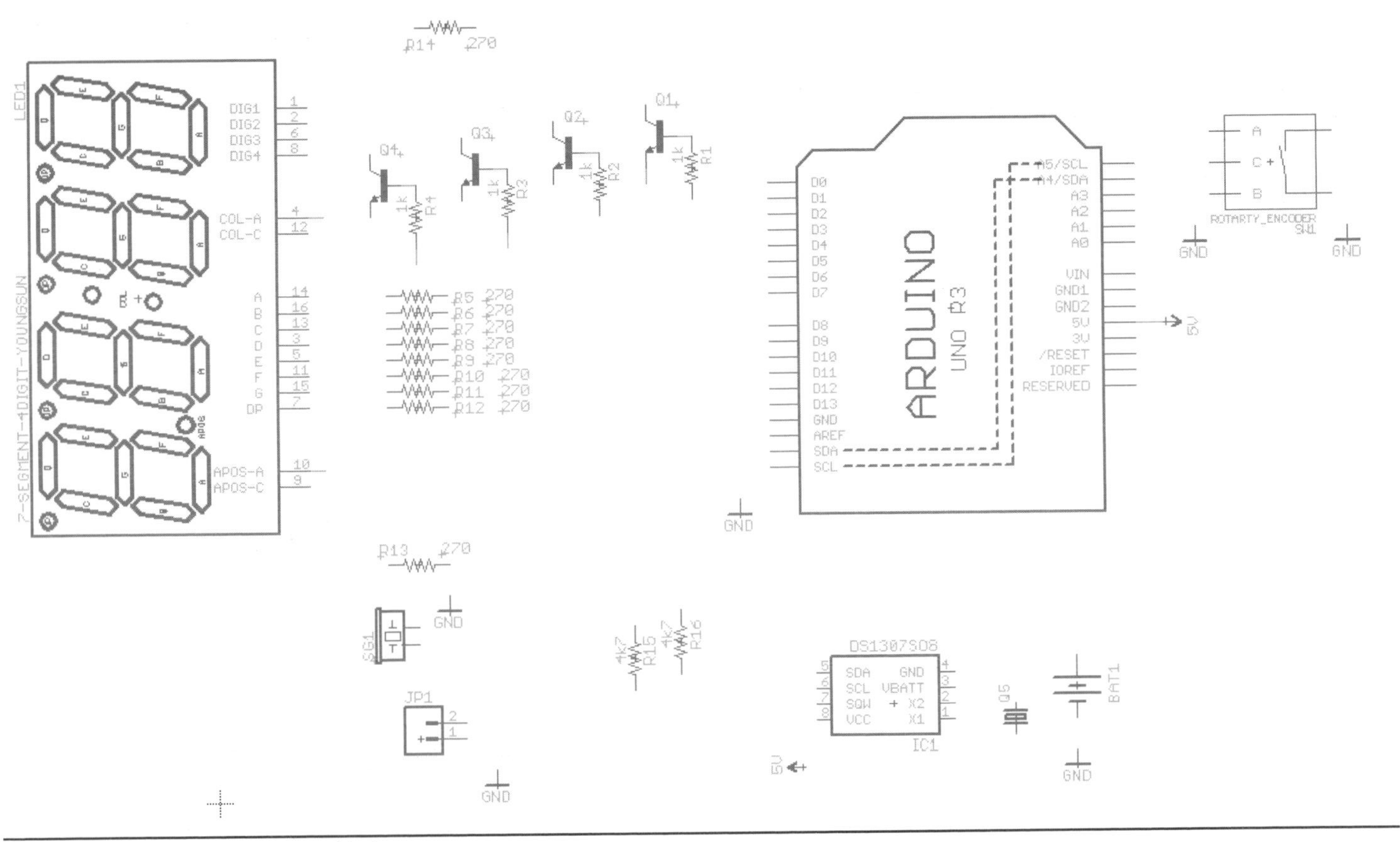

FIGURE 8-3 All the components added.

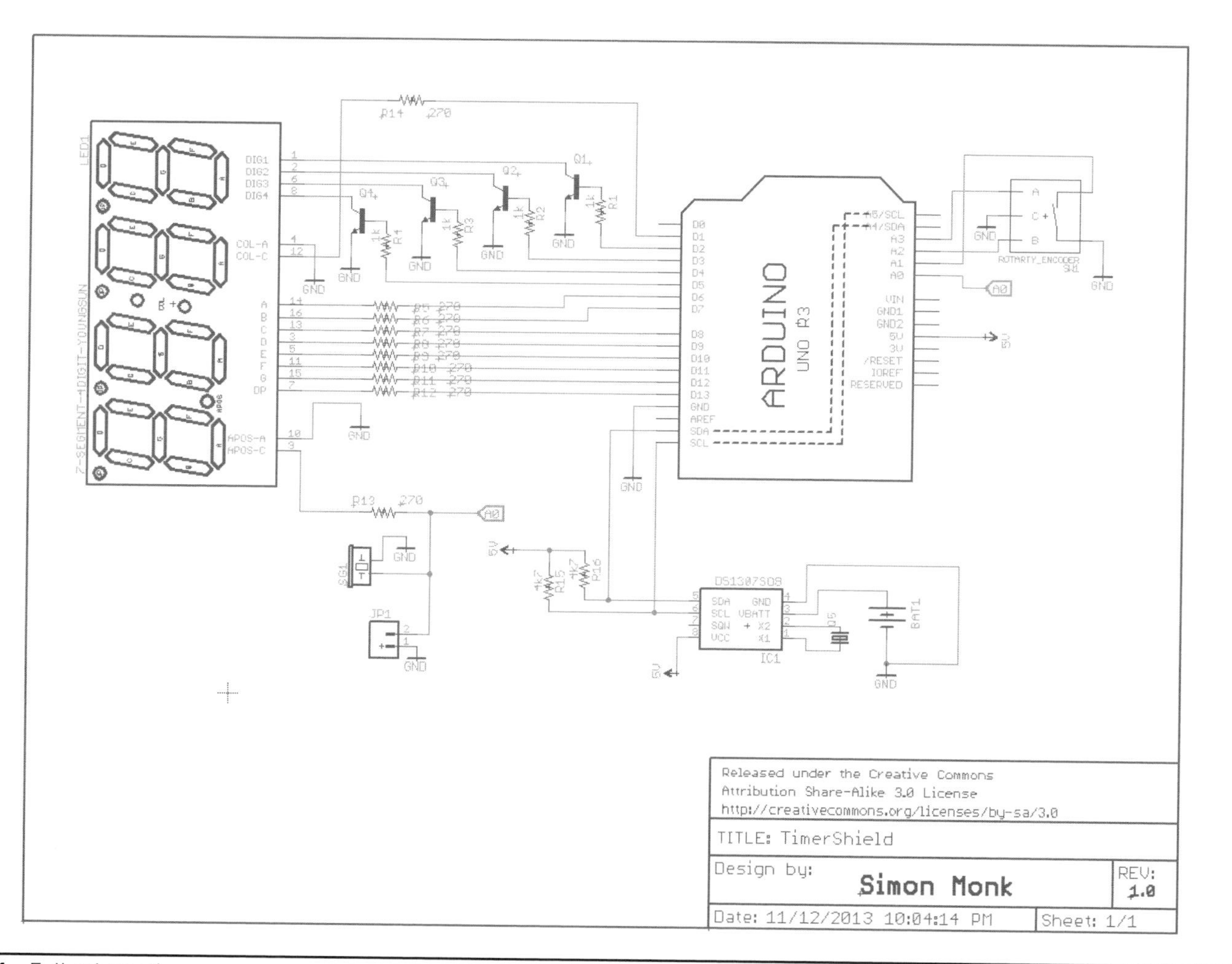

FIGURE 8-4 Full schematic.

Sometimes, in a schematic, it is not practical to connect both ends of a net with a line. For example, in Figure 8-4, the link between the Arduino pin A0 and the buzzer, JP1, and the apostrophe cathode are a long way apart. I could have drawn a line between the two areas of the schematic, but instead I have used the Label tool. To do this, you need to

- Make both ends of the net be the same net by setting them to the same name with the Name tool. I called them A0 after the Arduino pin.
- Make a short length of net just going to a blank area on the schematic, where we are going to add a label. Do this for both ends of the net.
- Select the Label tool, and then click on the end of the net segments you have just made to add the labels.

We should also set up a couple of net classes to set desired track widths when we come to run the autorouter. Open the net class editor from the Edit → Netclasses menu item, and set the default netclass width to be 12 mils. Add another netclass (Nr 1) called "Power," and set the width to be 24 mils (Figure 8-5).

This is by no means essential, but given that if all the segements of one digit are lit, there could be over 100 mA flowing through the collector-emitter path of the transistors. Thus, to change the net class of the four nets that connect the collector of the transistor to DIG1 through DIG4 on the display, you also should set any GND net path to be of class "Power" too. The easy way to change these netclasses is to use the Change command, select "Class" and then "1 Power," and then click on the nets you need to change.

Having drawn the schematic, it's time to run the electric rule checker (ERC). The result of this is shown in Figure 8-6.

At first sight, it seems like there are some things to worry about. But actually there aren't. The errors referring to power supplies all spring from unmade connections to power on the Arduino. Because the Arduino is providing power to the shield, not the other way around, we can safely ignore these. None of the warnings are relevant either.

Nr	Name	Width	Drill	Clearance
0	default	12mil	0mil	0mil
1	power	24mil	0mil	0mil

FIGURE 8-5 Adding a netclass.

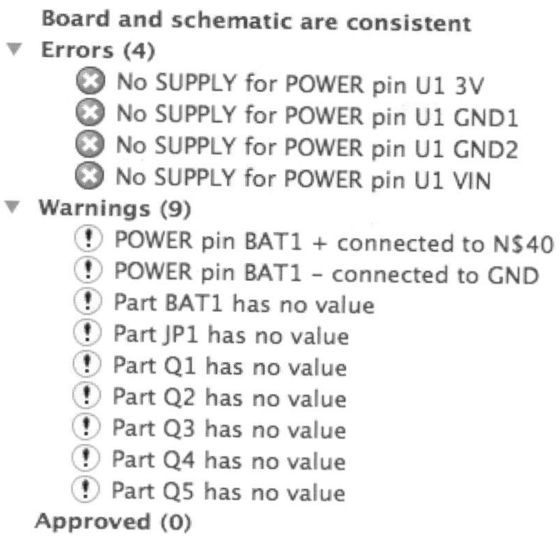

Board and schematic are consistent
▼ Errors (4)
No SUPPLY for POWER pin U1 3V
No SUPPLY for POWER pin U1 GND1
No SUPPLY for POWER pin U1 GND2
No SUPPLY for POWER pin U1 VIN
▼ Warnings (9)
POWER pin BAT1 + connected to N$40
POWER pin BAT1 - connected to GND
Part BAT1 has no value
Part JP1 has no value
Part Q1 has no value
Part Q2 has no value
Part Q3 has no value
Part Q4 has no value
Part Q5 has no value
Approved (0)

FIGURE 8-6 ERC results.

PCB Layout

When you first select the menu option "Switch to Board," the board layout that is subsequently generated will look something like Figure 8-7.

As you can see, as well as the shield outline in the "Dimension" layer, there is also a default rectangular board area with the layout origin at its bottom-left corner. This rectangular board can be deleted and the Arduino shield board moved so that the origin is at its bottom-left corner.

We can now start dragging the components onto the board, starting with the display and rotary encoder, which we want to be fairly central to the board. We can also position the battery holder over to the right edge of the board, rotating it so that the battery can be easily replaced.

The same argument applies to the screw terminal, which is located on the left of the board where the other Arduino leads for USB and power are attached so that all the leads are on one side and easy to access.

While giving the key components their initial positions, we also need to consider what is going on beneath the board. For example, the Arduino Uno has a metal USB socket in the top-left corner (see Figure 8-1). So we do not want any through-hole leads immediately above that, which might make contact with the metal of the socket. The same applies to the center of the right-hand edge of the shield, where there is a six-pin ICSP header poking up from the Arduino. This is another reason why it is a good location for the surface-mount battery holder to be mounted on the top side.

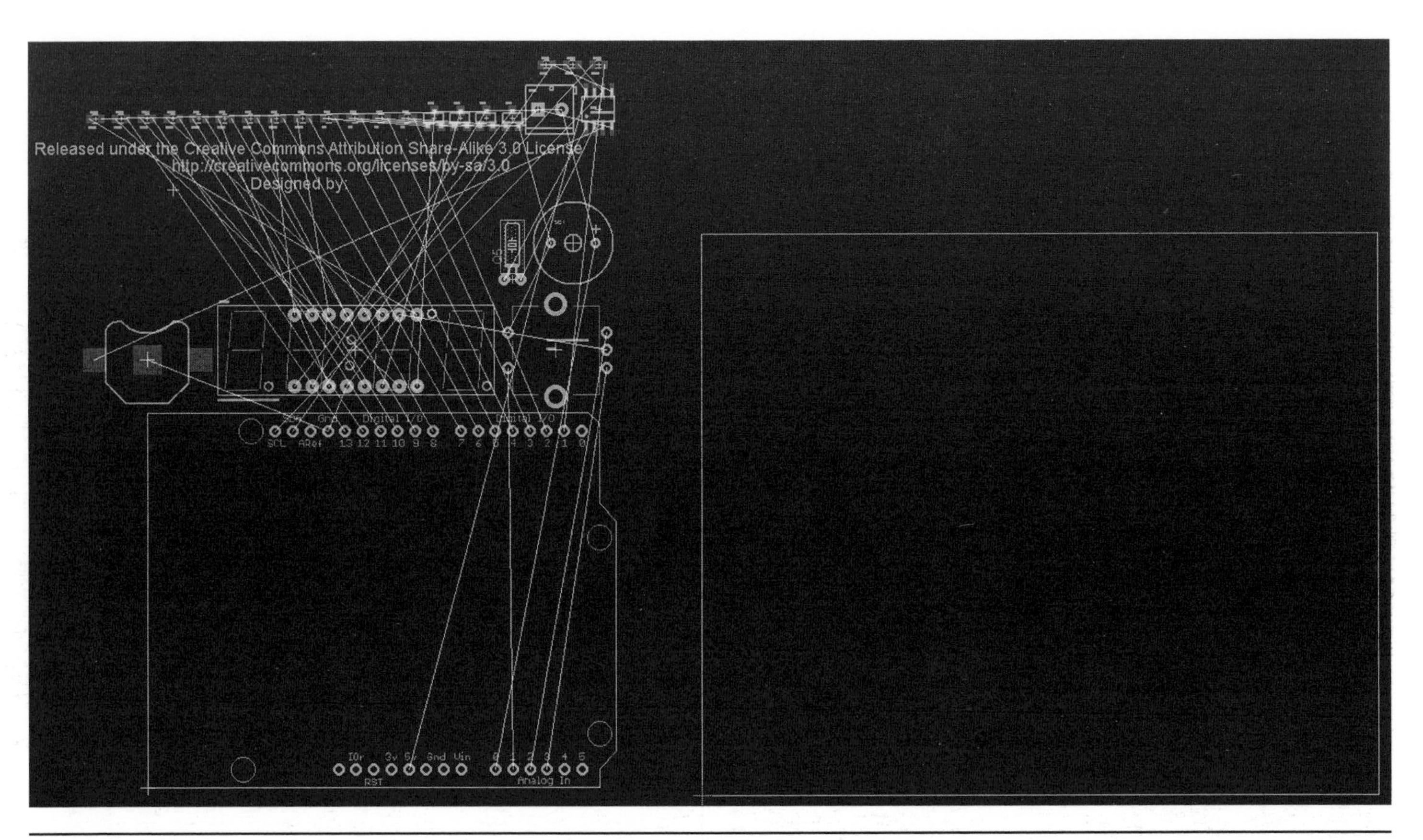

FIGURE 8-7 Initial layout editor mess.

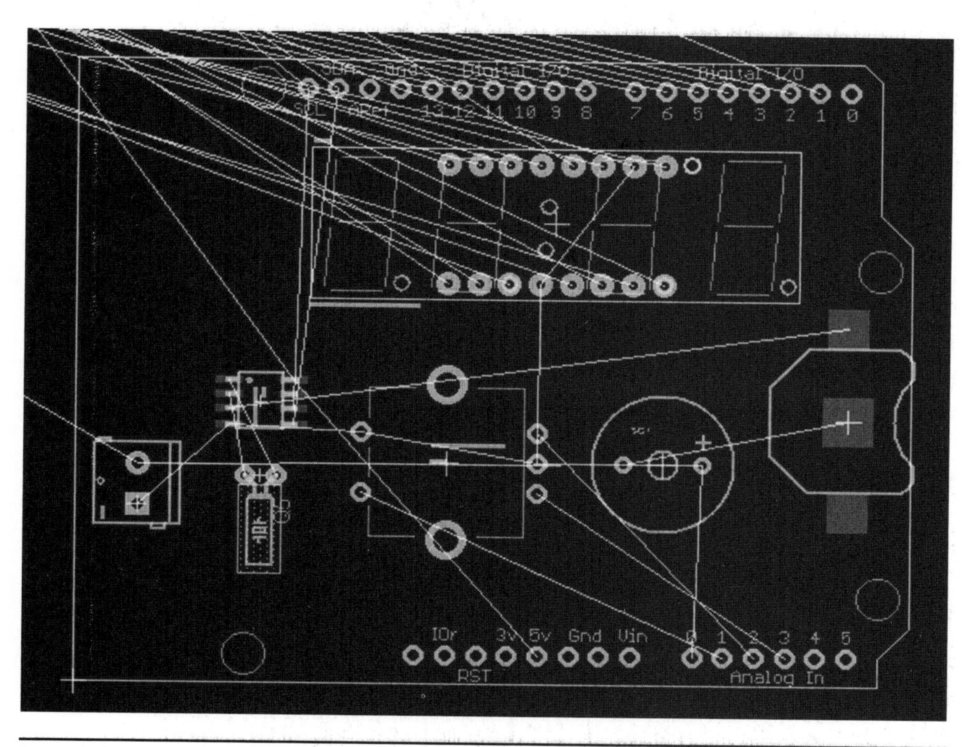

Figure 8-8 Key components positioned on the board.

With the key components positioned, the board will look something like Figure 8-8.

We now need to drag on all the transistors and resistors. Because most of the "Digital IO" pins at the top of the board will have a resistor attached to them, drag over the appropriate resistors, positioning them under their input-output (IO) pin and rotating them as appropriate.

The transistors can be placed within the outline for the display. Because the display stands a little off the PCB, they can be on the top layer.

Figure 8-9 shows all the components on the board. Note that the air wires around the transistors seemed to be crossing over rather a lot, so they were rotated. It was also noticed that the buzzer has a polarity and that the positive (+) connection was currently to ground. The pins were swapped over in the schematic.

We will add a ground plane to both the bottom and top layers. Remember, you do this using the Polygon tool, first setting the layer to "Bottom." To get this to follow the shape of the board, you will also need to change the line style to a simple straight line rather than the normal orthogonal (right-angle) line style. This is next to the dropdown layer choice, once the Polygon command has been

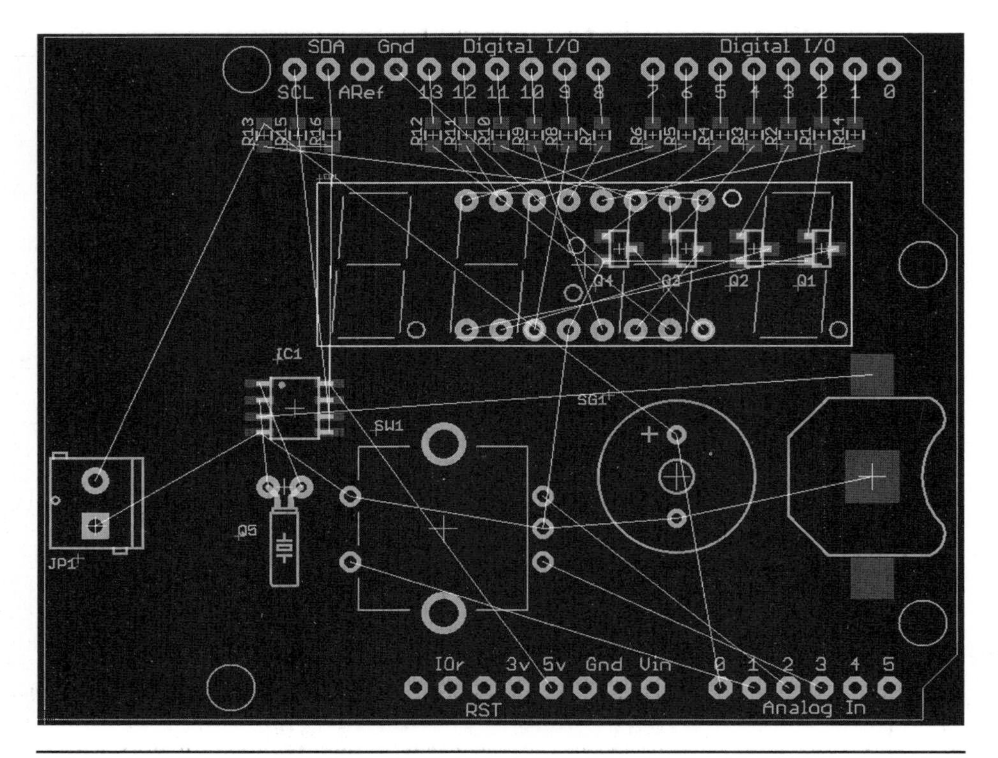

Figure 8-9 All the components on the shield.

selected. Having drawn the polygon, use the Name command to set the name to be GND, and then hit the Ratsnest command to see the effect. Repeat this whole process but on the top layer. Because there are through-hole connections such as the GND connection of the screw terminal, there will automatically be a link between the two ground planes.

Run the autorouter, changing the routing grid to 25 mils before running. It should route 100 percent and produce a layout like that in Figure 8-10.

We need to tidy up all the text on the board. We want any text that is to appear to be on the tPlace layer. So use the Smash command to separate all the labels from their parts, and then use the Change command to set the layer of the text that you want to keep to be tPlace.

Now use the Change command to make all the labels a Vector font and then set their size to 0.032.

Because this is an OSH project, we should also put the OSH logo on the board. This is a rare occasion where we can add a part to the Layout Editor rather than the Schematic Editor. Select the Add command, and then search for *OSH* in the component library and select the icon OSH-LOGO-L.

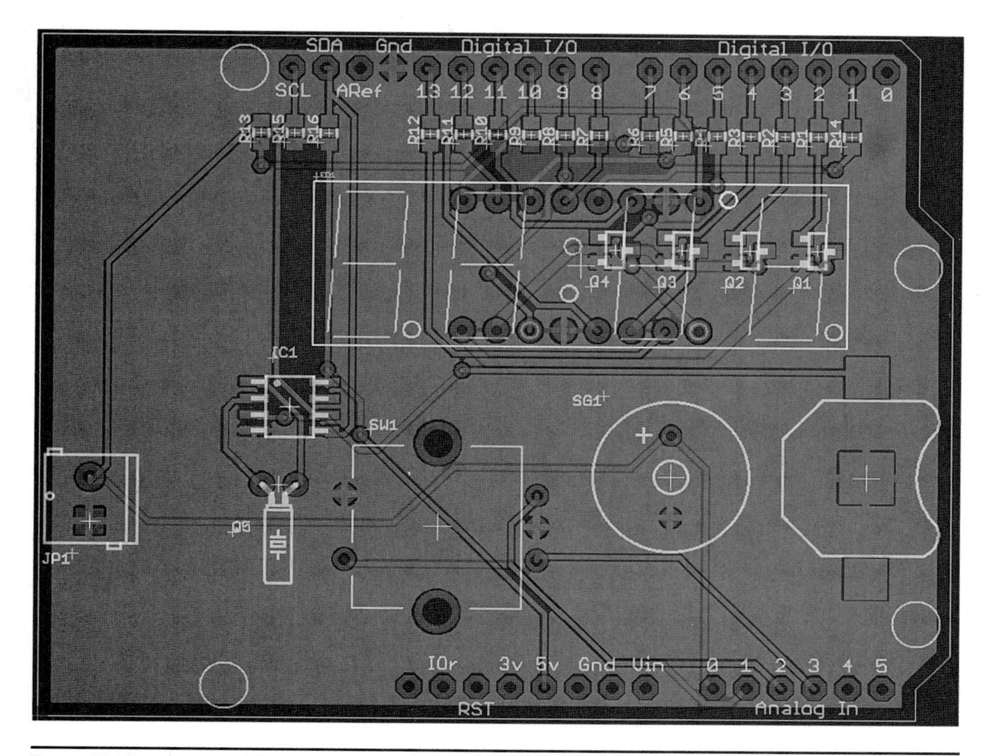

FIGURE 8-10 A routed layout.

Also check for anything that is in the "tDocu" layer that you want to appear on the silk screen of the finished board. Because some parts such as the buzzer have their outline on the "tDocu" layer, we can just add the "tDocu" layer to the CAM job before we run it. We can also add some text for a URL next to the logo using the Text tool.

One of the problems with using an Arduino shield is that of documentation—knowing which of the Arduino pins are used and for what. A great place for this documentation is on the board itself, perhaps in the form of a little table. To draw this, it helps to set the grid to, say, 0.01 mil. Remember to set the layer of the lines to be "tPlace," and a line width of about 0.05 mil will be about right.

We can check what will appear on the silk screen of the board by hiding all the layers except "tPlace" and "tDocu." It is also useful to show "Pads," "Vias," and "Dimension" (Figure 8-11). The small cross marks next to each label will not appear on the final silk screen.

With a couple more labels added for JP1, the final board layout will look something like Figure 8-12.

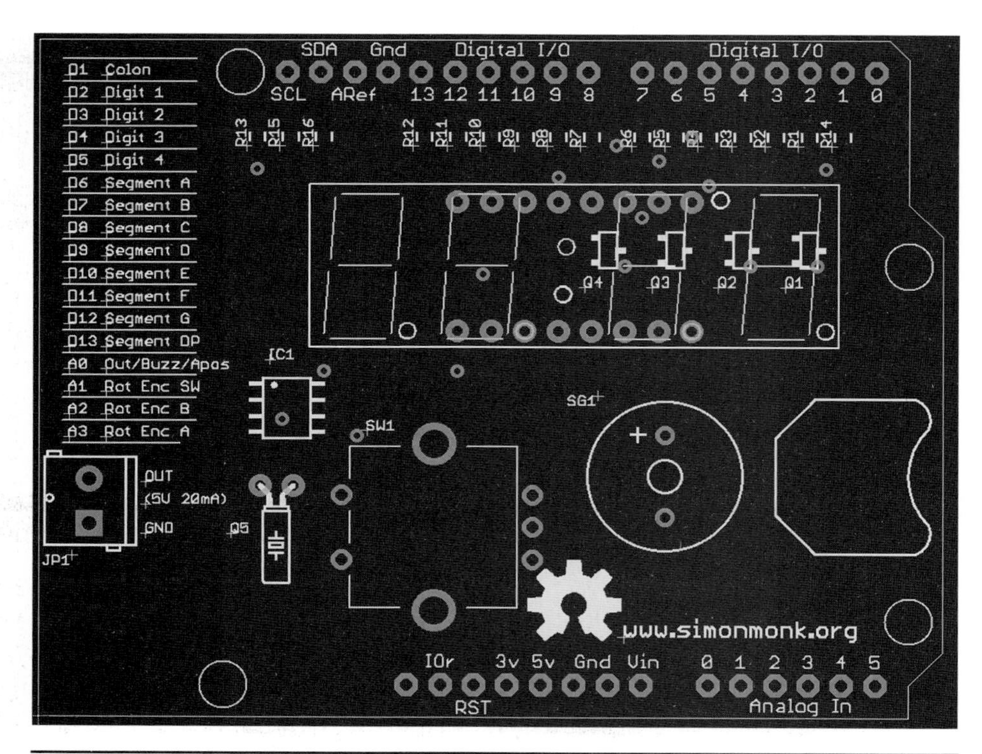

FIGURE 8-11 Checking the silk-screen appearance.

Fabrication

To generate the Gerber files for fabrication, we will use the Sparkfun CAM job that we saw in Chapter 6. However, having opened the CAM job, go to the "Top Silk Screen" tab and select the "tDocu" layer so that both the "tPlace" and "tDocu" layers are selected (Figure 8-13).

The bounding rectangle for a shield board such as this is 2.7 × 2.1 in. (68 × 53 mm). If you shop around, you should be able to buy five boards for around $20.

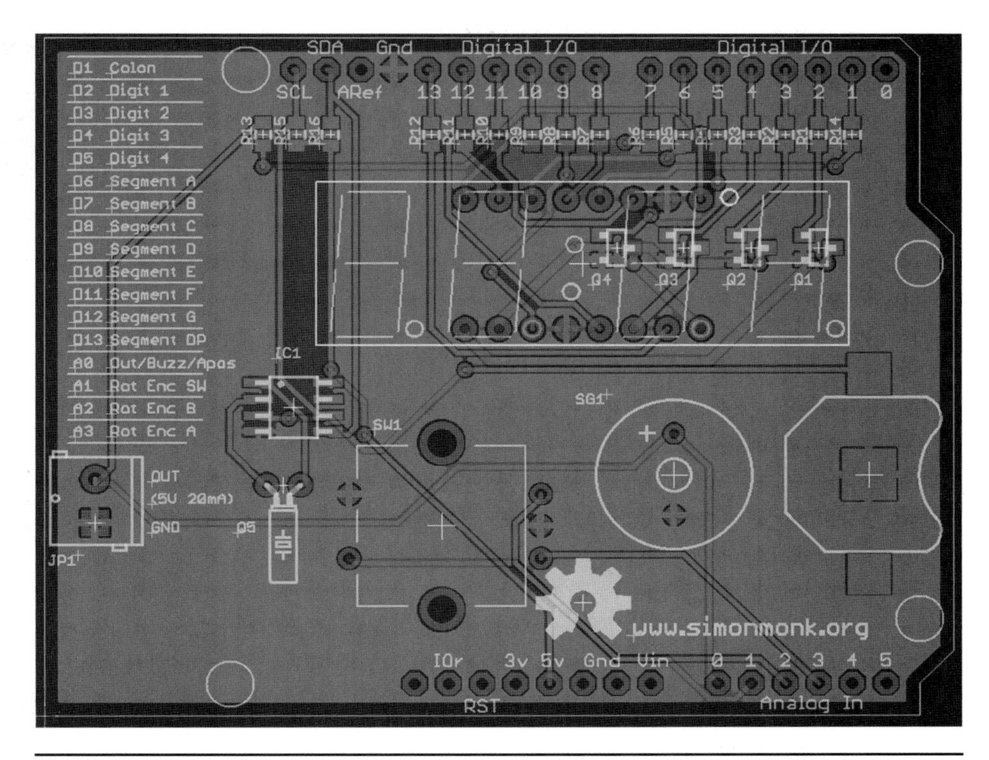

Figure 8-12 Final board layout.

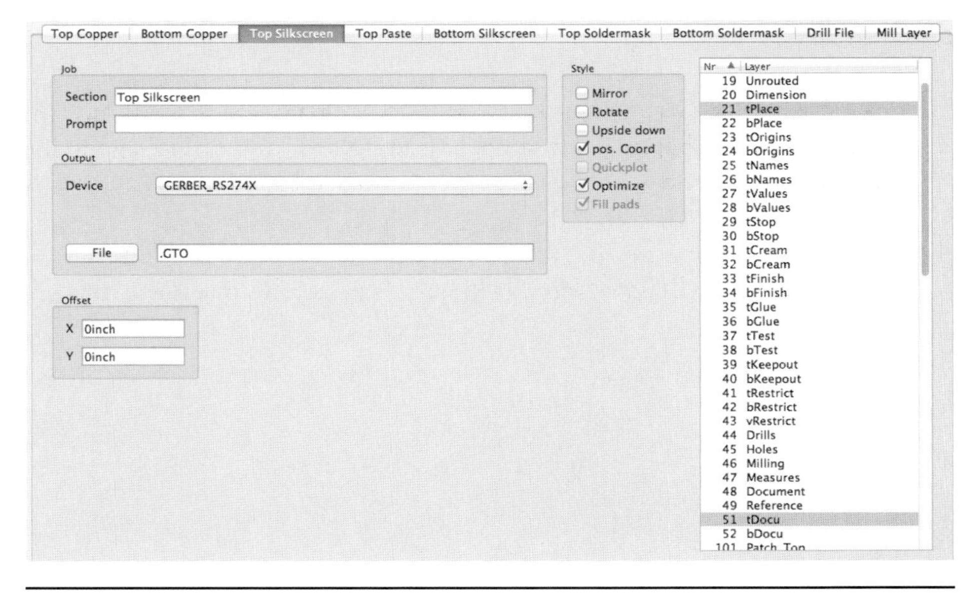

Figure 8-13 Selecting the "tDocu" layer in the CAM job.

Summary

If you are interested in making this shield, be aware that it is as yet untested. The LED display used has a manufacturer's ID of TDCG1060M and is available from Mouser and Digikey.

In this chapter we have looked at designing an Arduino shield using EAGLE. In Chapter 9, we will look at another example project, this time building an interface board for the Raspberry Pi single-board computer.

CHAPTER 9

A Raspberry Pi Expansion Board

In this chapter, we will look in detail at another of the author's open-source hardware (OSH) projects, specifically, version two of the RaspiRobot board (www.raspirobot.com). This board (Figure 9-1) is designed to allow a Raspberry Pi single-board computer to be powered by batteries and also to control two direct-current (dc) motors. The board has a number of interfaces including to an HC-SR-04 ultrasonic range finder, a 5V I2C interface designed to easily accommodate displays and other I2C devices, and a pair of open-collector outputs.

In following this design, you will learn how to develop your own add-on boards for the Raspberry Pi.

Design Considerations

The board is interesting from a design point of view because it uses a switched-mode power-supply chip that has very specific routing requirements. The board also uses a mixture of both through-hole and surface-mount devices. Figure 9-2 shows a block diagram for the design.

The round objects are external to the RaspiRobot board, and the rectangular blocks are part of the board. This type of diagram can be very useful for setting the context of the project and understanding how it will interface with other parts of the system.

Figure 9-1 The RaspiRobot board version 2.

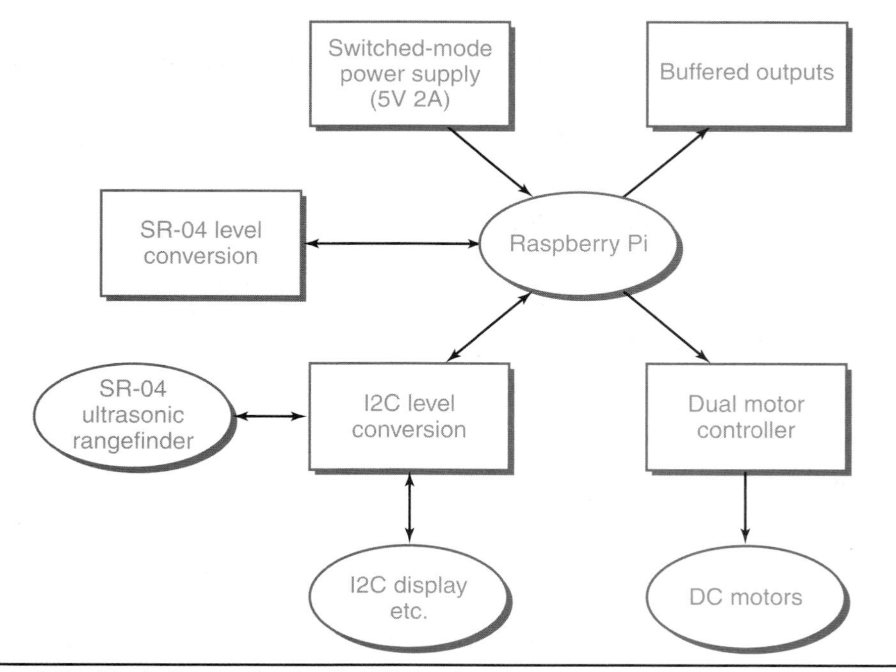

Figure 9-2 Block diagram for the RaspiRobot board.

The Schematic

There is nothing particularly interesting about the schematic (Figure 9-3), so we will not dwell on it for long.

One quite common pattern in the schematic design is worth mentioning, and that is the separating out of the power-supply section of the board. This has the components for the power supply itself and the decoupling capacitors (C3 and C4) on their own. Separating the power-supply section of the board like this is a lot clearer than adding nets as lines between all the components.

The Raspberry Pi, to which the board will be attached, is represented in the schematic as a 2- by 13-in. socket header.

The Board

The board was designed using a mixture of manual and automated layout. Figure 9-4 shows the initial placement of the components.

In considering the board design, there are two areas where it is worth some manual layout before we unleash the autorouter. These are

- The critical layout considerations for the power-supply chip IC2 and its associated components
- The power supply and outputs of the motor controller chip IC3

Laying Out the Power Supply

As you can see, the power-supply components take almost a third of the right-hand side of the board. The switched-mode power supply is based around the LM2596S chip. The datasheet for this integrated circuit (IC) indicates that there are certain paths between its associated components that need to be as short and low resistance as possible; otherwise, instability will result. There are also some fairly large currents flowing here. Figure 9-5 shows the relevant part of the datasheet indicating the paths that need to be kept as short as possible.

As the note in the datasheet states, a ground-plane construction is ideal, and we will use this for much of the power-supply area, with some short and thick tracks for the areas not connected by the ground plane.

We will use a ground plane over the whole bottom layer and a second ground plane around IC2 D1, C1, and C2. The inductor does not have a connection to

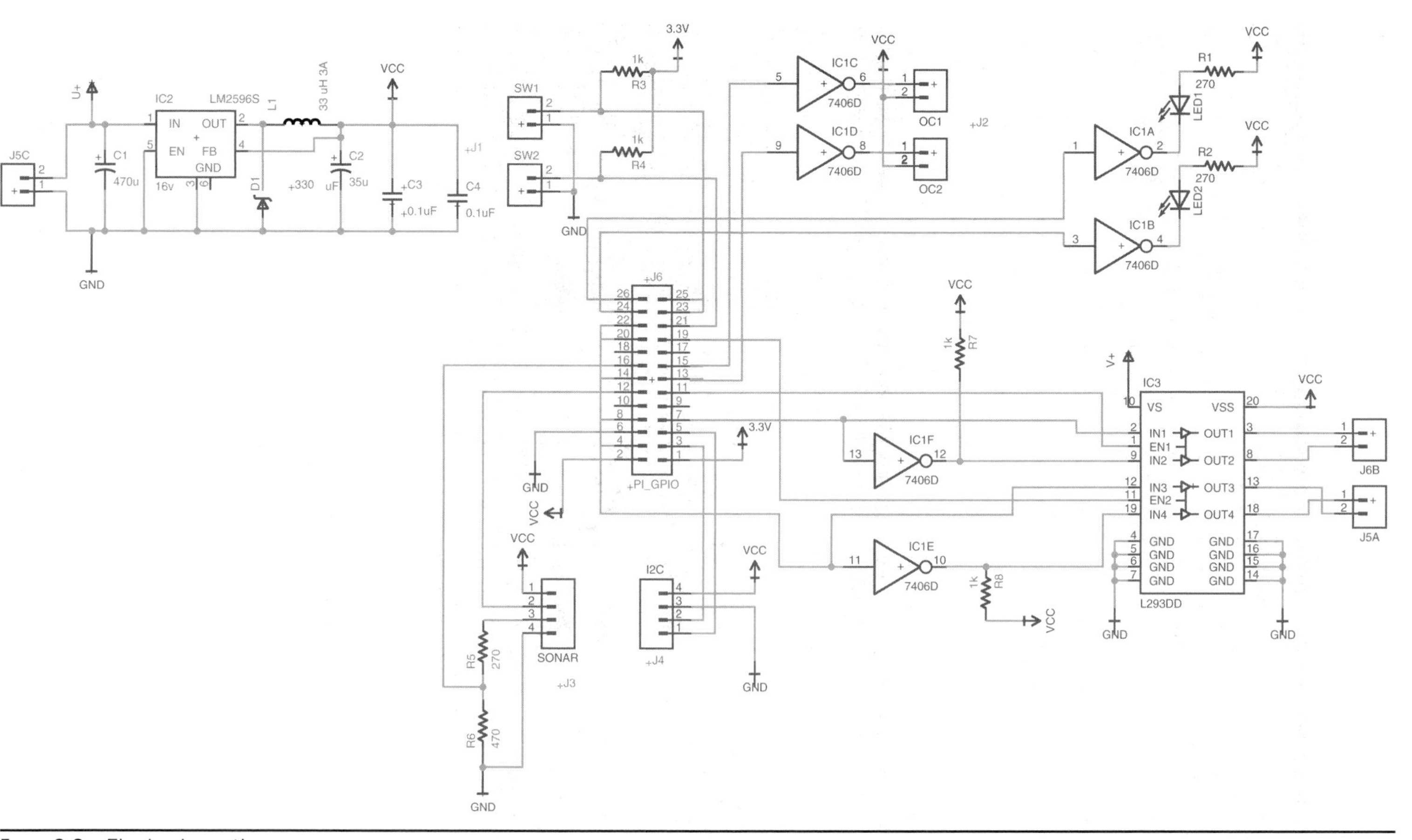

Figure 9-3 Final schematic.

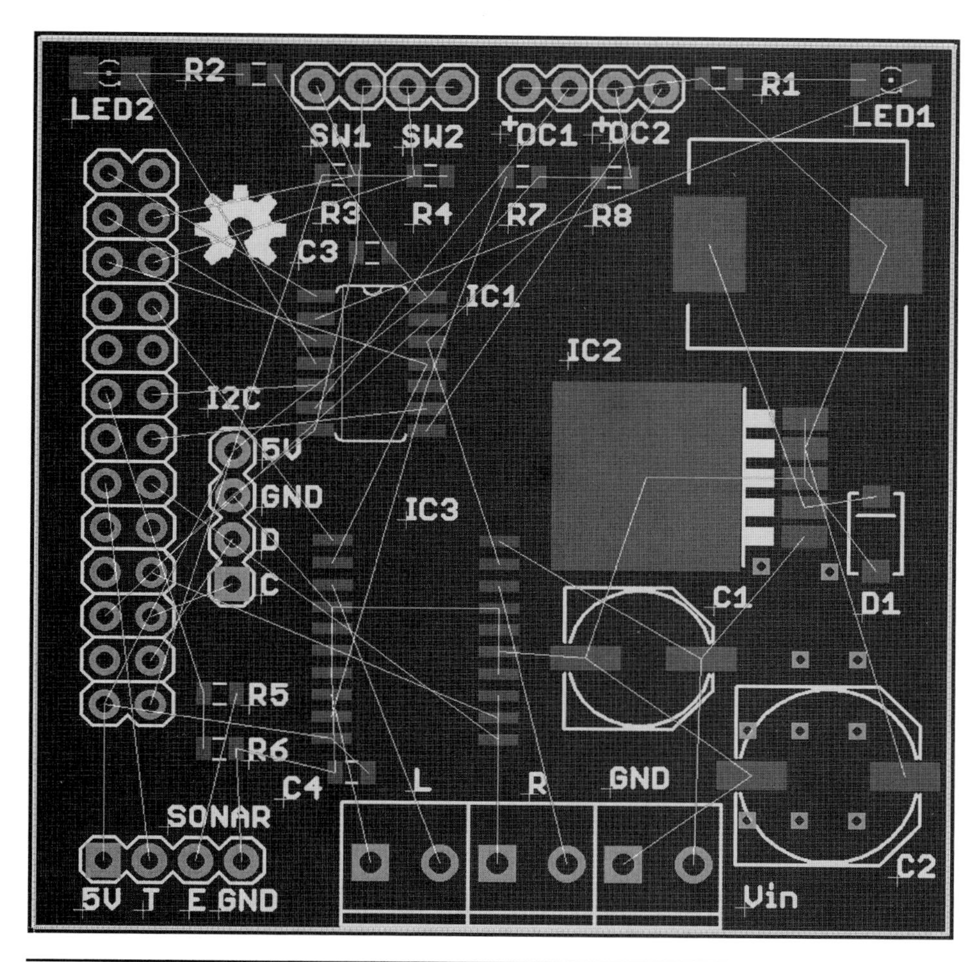

FIGURE 9-4 Initial component placement.

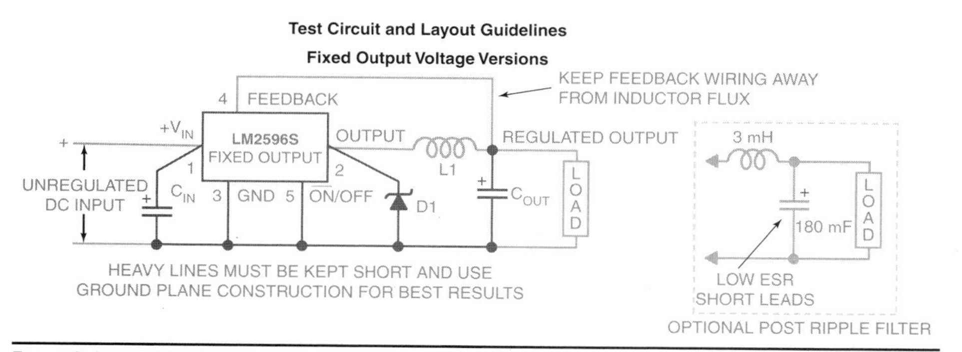

FIGURE 9-5 Critical connections on the LM2596S.

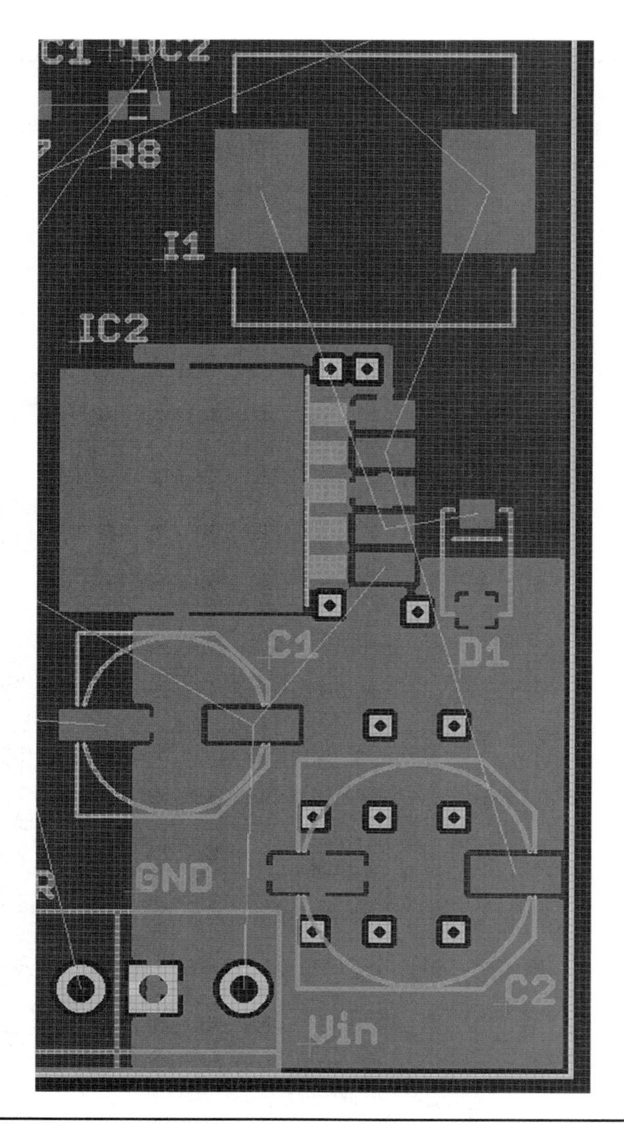

Figure 9-6 Adding ground planes.

GND because it will be connected to pins of IC2, which must be on the top layer; thus there is little point in extending the ground plane up to L1. Figure 9-6 shows the board with the two ground planes added. Note how I have added a grid of vias to link the two ground planes.

This figure indicates that the connections between L1 and IC2 are not critical, but common sense dictates that these tracks will be carrying significant current, so let's start by routing them with some fairly thick tracks (40 mils). The connection between L1 and D1 is also marked as critical, so let's do the same for that connection. This last net also connects to pin 2 of IC2. This is handily close to D1, but if we used 40 mils, we would be getting close to neighboring pins of IC2, so for this short track the width was dropped to 32 mils.

While we are dealing with the high-current tracks, let's also add in the track between Vin, C1, and pin 1 of IC2. Again, let's make these in 40-mil width. Remember to keep clicking the Ratsnest command when routing through the top ground plane. The result of these routings is shown in Figure 9-7.

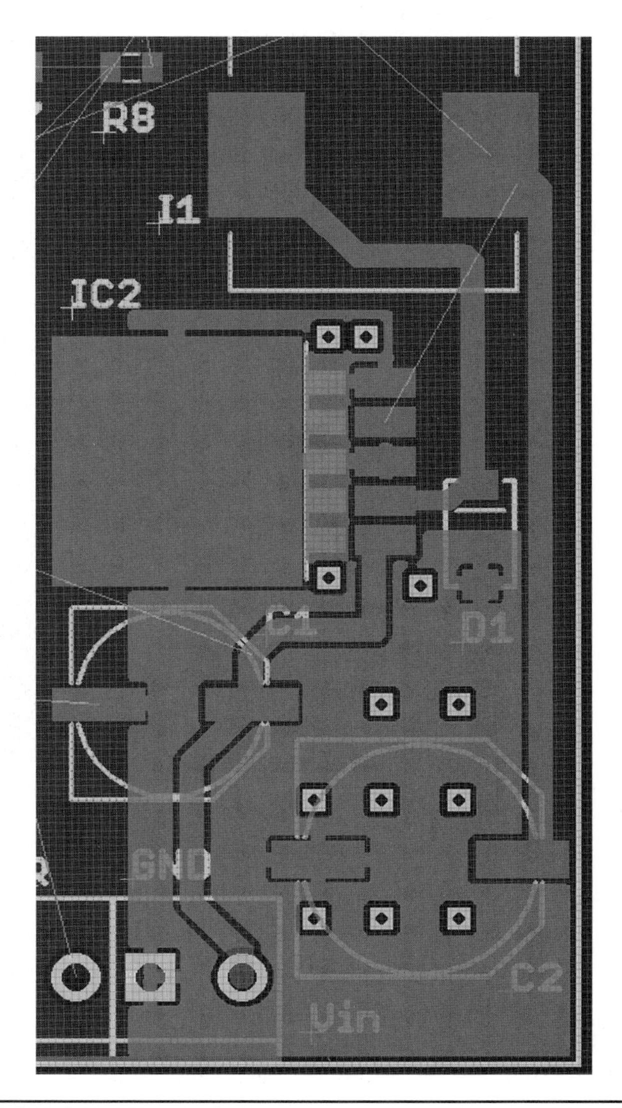

FIGURE 9-7 Routing the power supply.

Referring again to Figure 9-5, pin 4 of IC2 is a feedback signal. It measures the voltage being output. This will not be carrying any significant current, but the datasheet indicates that it should be kept away from the inductor. This track can just travel out horizontally from pin 4 of IC2. It does not need to be thick, but it does need to get past the track on the top layer from C2 to L1. We can do this by hopping under it using a pair of vias.

When routing from one layer to another, do not place the vias first. They will be placed automatically for you as you switch layers. This is illustrated in Figure 9-8.

First, set the width to 10 mils, and start making the track from IC2 pin 4 on the top layer (Figure 9-8*a*). When you get halfway to the vertical track that we want to go under, left-click the mouse. This is where the via will go when we switch layers. Thus, without changing away from the Routing tool, move the mouse up to the layer dropdown and select the bottom layer. Move the track on a little further to the right. You will see that it is now routing on the bottom layer (Figure 9-8*b*). The via has not yet appeared, but it will soon.

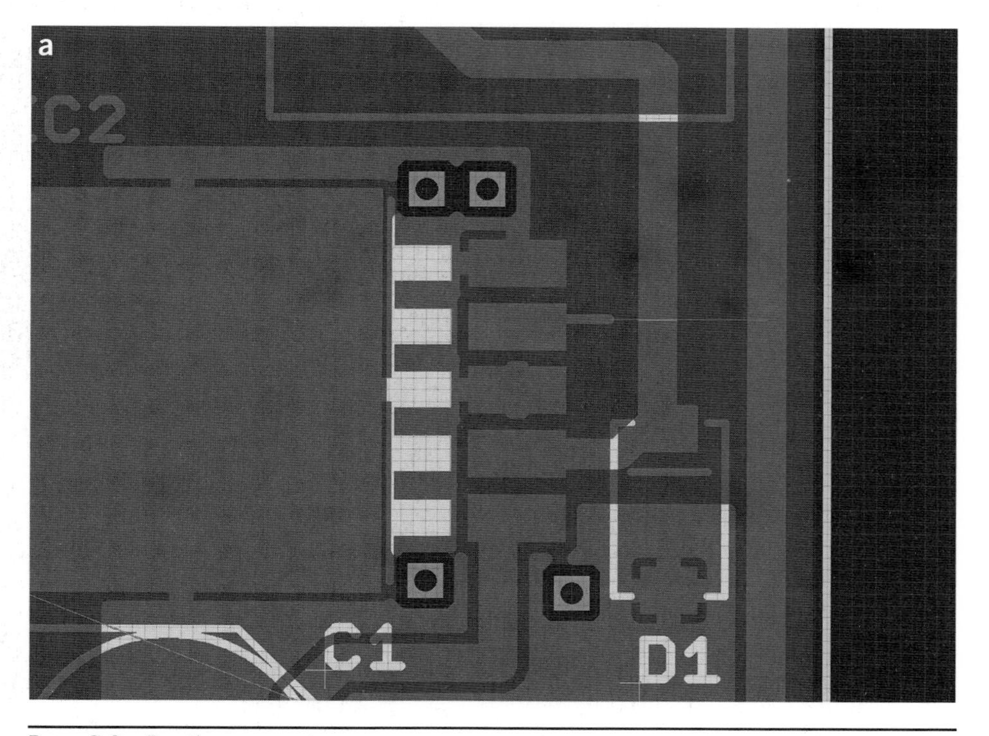

Figure 9-8 Routing across layers.

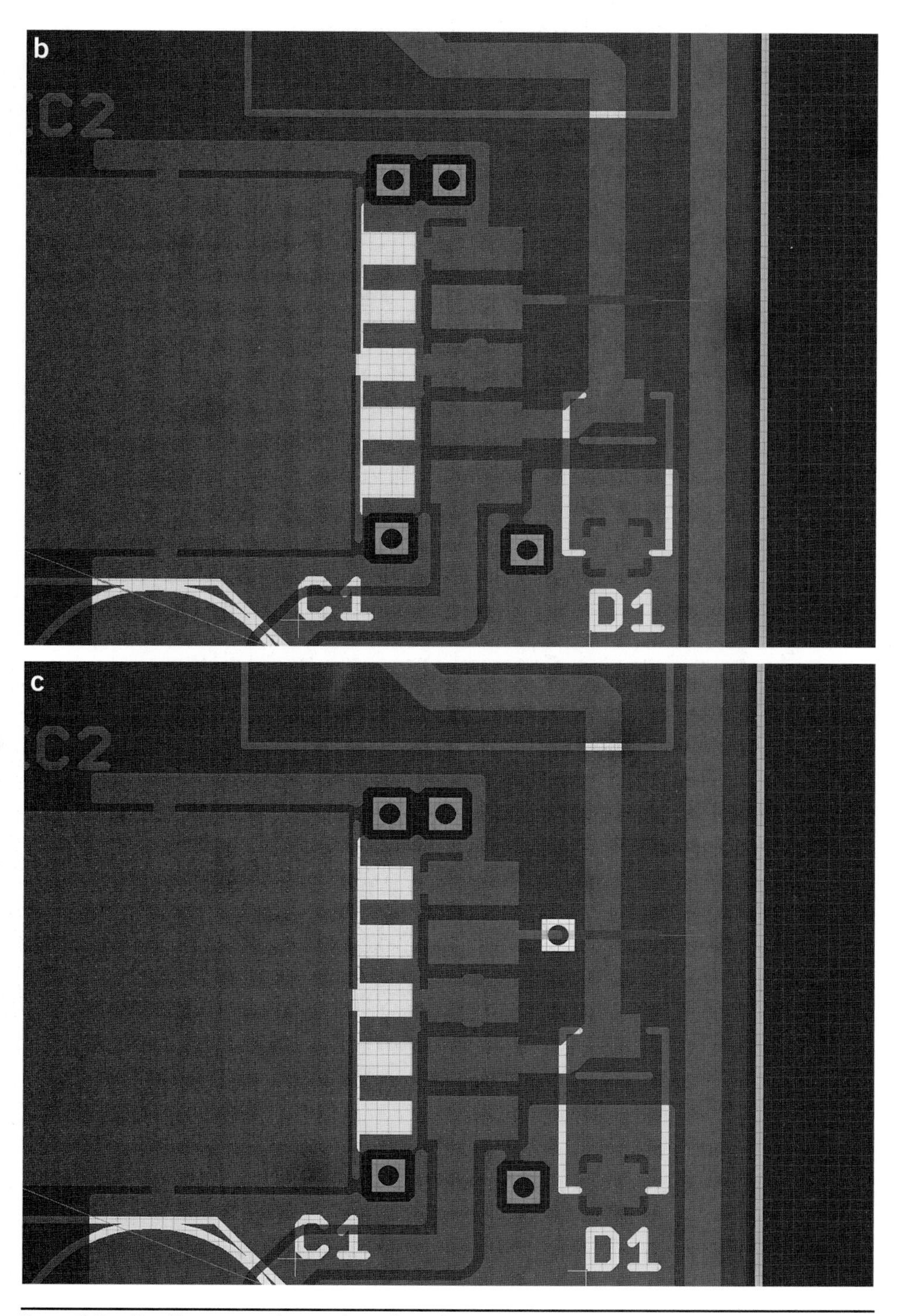

Figure 9-8 Routing across layers (*continued*).

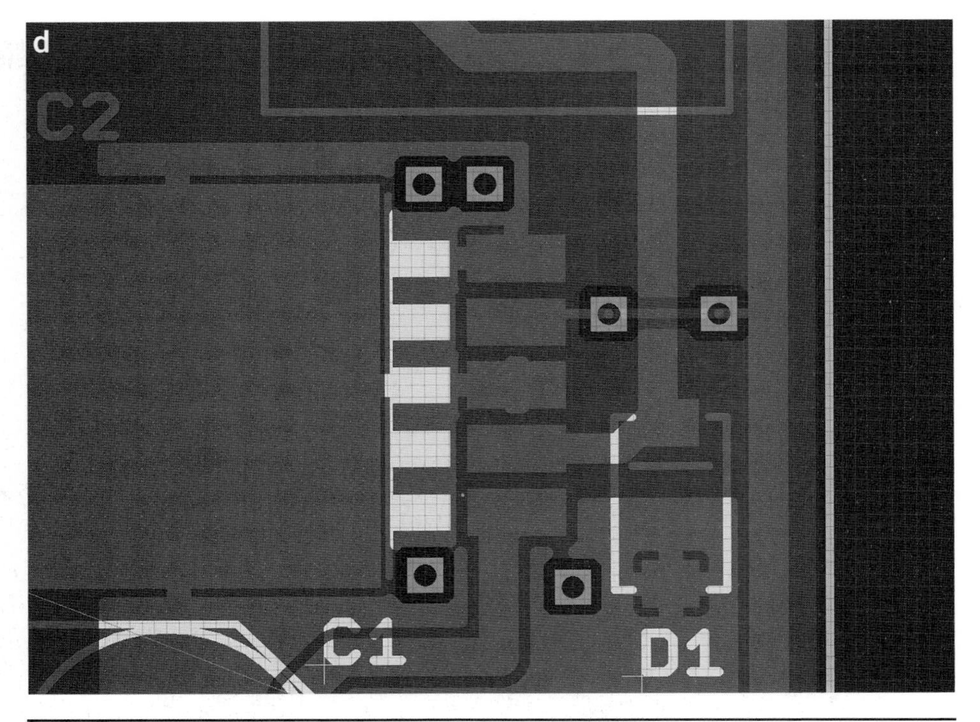

Figure 9-8 Routing across layers (*continued*).

Almost immediately we will need to swap layers again, so left-click the mouse where you want the via, and then select the top layer again. At this point, the first via will appear (Figure 9-8*c*). Finally, complete the track, routing it to the right-most vertical track, and both vias will be visible. Click on "Ratsnest" to make a gap around the track on the bottom layer so that it is not merged with the bottom ground plane (Figure 9-8*d*).

Laying Out the Power and Motor Tracks

The other area of the board that is worth manually routing concerns the other high-current tracks, that is, the tracks from IC3 to the screw terminals that provide power to the motors and also the regulated 5-V line that provides power to the Raspberry Pi. Again, the width is chosen to be as wide as can be easily accommodated. In some places, this width is reduced midway along the track as it nears a pad. Figure 9-9 shows the final board with these air wires routed and the remainder of the board routed using the autorouter.

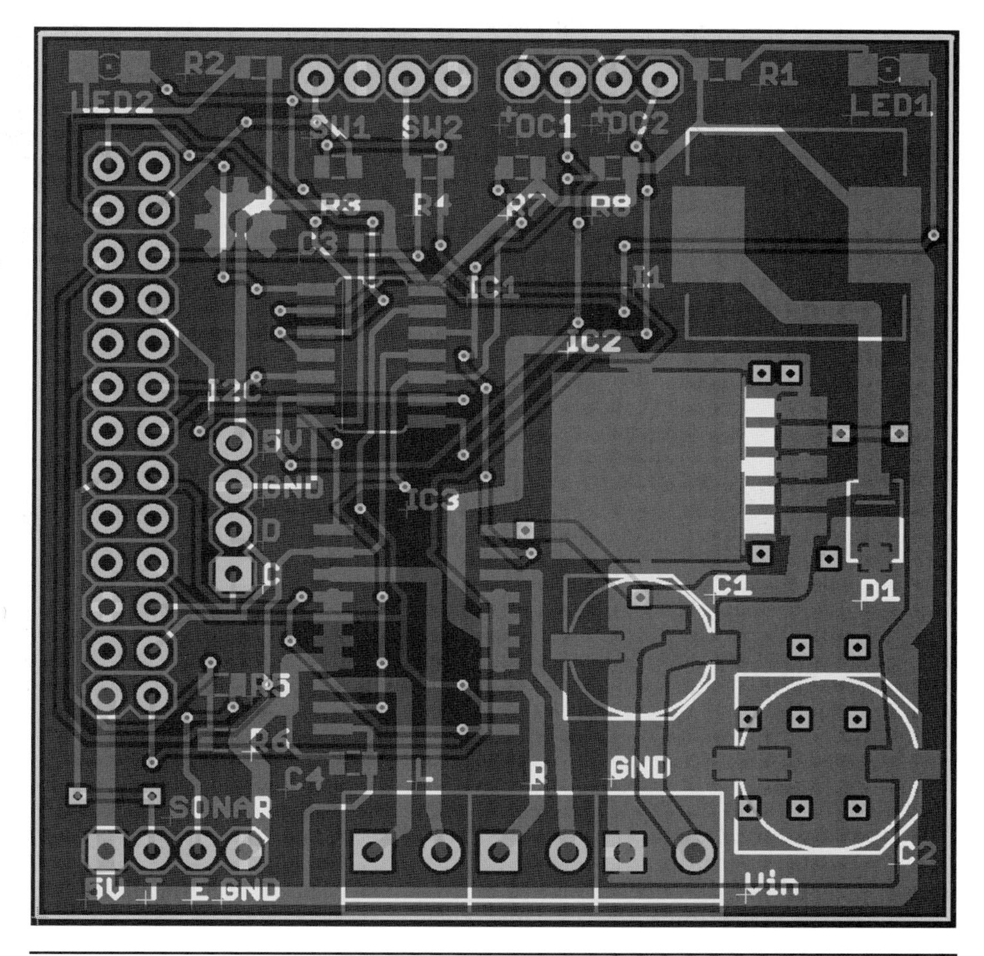

FIGURE 9-9 Final board layout.

Summary

In this second example chapter we have learned about routing with a mixture of manual routing and the autorouter. In Chapter 10, we will look at how you can interact with EAGLE using text commands and then extend this to automate what would otherwise be tedious and repetitive tasks using scripts and user-language programs (ULPs).

CHAPTER 10

Commands, Scripts, and User-Language Programs

Sometimes, when designing an electronic project in EAGLE, you will come across activities that are repetitive or that you seem to need to do for every project you start. It would be nice to be able to automate such activities and save yourself some time.

Fortunately for us, EAGLE includes technology to do just such automations. *Scripts* are simple lists of text commands to be invoked and are useful enough. However, EAGLE also includes a *user-language program* (ULP) language. This is a fully featured programming language that allows you to branch off and do different things under different circumstances or repeat certain instructions a number of times in a loop. For example, later we will look at how you can use a ULP to automatically smash all the parts on a schematic.

Ultimately, whether you use a ULP language or a simpler script, the end result will be the invocation of commands, and it is here that we will start.

Commands

You may have noticed the command line beneath the toolbars. You can type commands here that do the same things as you can otherwise accomplish using your mouse. Figure 10-1 shows a command ready to be run.

Start a new project with a schematic, then enter the command `add R-US_VMTA55@rcl (0 0.1)`, and then hit "Return." You will see a resistor appear just above the origin. Let's now examine this command.

The command is Add, and this can be specified in uppercase or lowercase—it doesn't matter. You also can just use enough letters to make the command unique—in this case, just `a` would work fine.

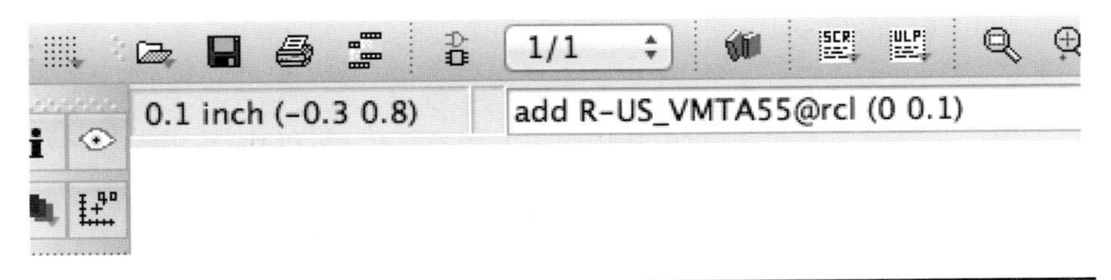

Figure 10-1 Entering commands.

The next section of the command line specifies what is to be added. In this case, it is the part `R-US_VMTA55` in the library `.rcl`. Following this are the coordinates where the part is to be placed.

Let's add another resistor just above the last one using the command

```
add R-US_VMTA55@rcl R0 (0 0.3)
```

You should now have two resistors on the schematic, and you can delete them by typing the following commands:

```
delete R1
delete R2
```

All the commands on the Command toolbar have text equivalents that can be typed into the command line. You can find full documentation on all these in the built-in help accessible from the Help option on the Windows menu. Open the section called "Editor Commands."

The command line is also available on the Board Editor. You also may have noticed the dropdown list on the right-hand side of the command line. This contains a list of previous commands used. When you select one of these, it will be copied into the command line, where you can edit it before running it. You also can access previous commands using the up-arrow on your keyboard.

Some people just prefer using the keyboard over a mouse and find this way of working natural, but the real power of these commands is in combining them into scripts and generating them from ULP languages.

Scripts

A *script* is just a list of commands held in a text file that you can then run. EAGLE includes a set of script files that are ready to use. You will find them in the `.scr` folder in your EAGLE installation folder.

Built-in Scripts

Let's try out one of these scripts. The one you are going to try is called `euro.scr`. You can browse all the scripts in the "Scripts" section of the Control Panel. When you select one of the scripts in the list (Figure 10-2), a description of the script is displayed on the right-hand side of the Control Panel.

If you double-click on the script in the Control Panel list, it will be opened in a text editor and will look something like this:

```
# Euro Format Board
#
# Draws the dimension lines of a euro format board (100mm x 160mm).


Grid mm 1 off;
Set Wire_Bend 0;
Layer Dimension;
Wire 0  (0 0) (160 100) (0 0);
Layer Top;
```

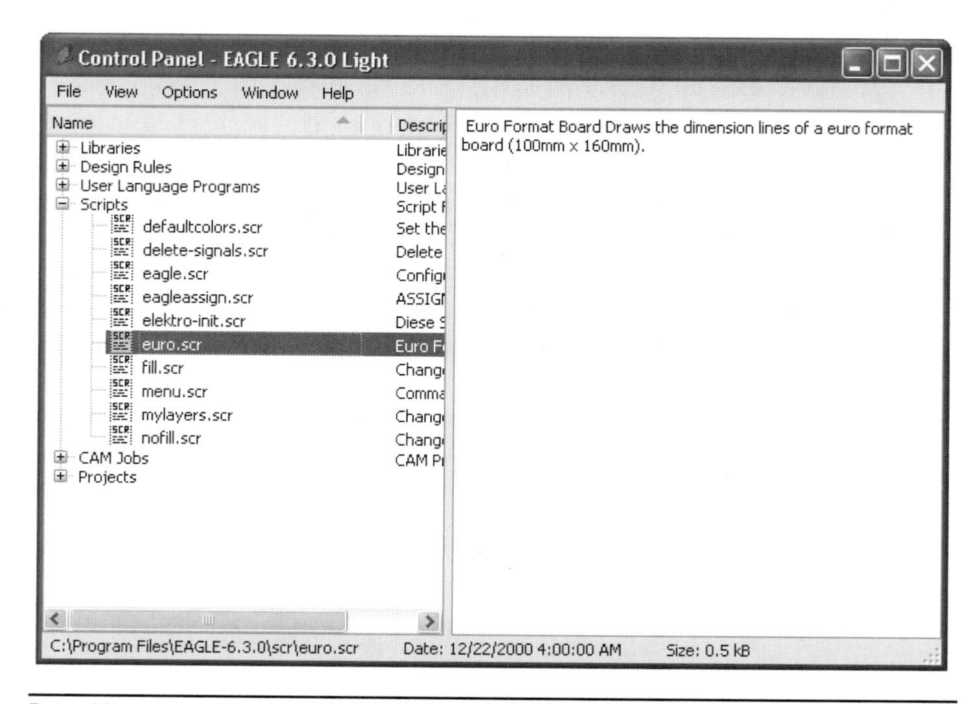

Figure 10-2 Selecting a script in the Control Panel.

```
wire 2  (20 -1) (-1 20);
wire  (140 -1) (161 20);
wire (20  101) (-1  80);
Set Wire_Bend 4;
wire (161 80) (140 101);
Layer Bottom;
Set Wire_Bend 0;
wire 2  (20 -1) (-1 20);
wire  (140 -1) (161 20);
wire (20  101) (-1  80);
Set Wire_Bend 4;
wire (161 80) (140 101);
Grid Last;
Window Fit;
```

The three lines at the top starting with a # are just comment lines that explain what the script does. It is a script to be run in the Board Editor that creates a Euroboard. Euroboard is a standard size of board sometimes also called a *Eurocard*.

Switch to the board layout (creating one if needed) for your example project, and delete the existing wire rectangle in the "Dimension" layer. To run the command, click on the small "SCR" icon on the toolbar. This opens a dialog window from which you can select the script `euro.scr`. When it has run, the board should look something like Figure 10-3.

Writing a Script

If you have a section of design that you find yourself repeating often, then you could make yourself a script that adds all the parts and connects them up. For example, it is pretty common to use a 7805 voltage regulator and its accompanying capacitors in a circuit. You can write a script that will add the necessary components and then connect them up. This, of course, will take place in the Schematic Editor.

Probably the simplest way to do this is to enter the commands one at a time into the command line and then transfer them a line at a time into a Text Editor. Periodically, you can save the script file, delete everything off the schematic, and then continue adding commands until it does what you want.

To create a new script, select the option "New Script" from the File menu in the Control Panel, as shown in Figure 10-4.

The logical place to start is with the 7805 itself. Wherever the assembly of voltage regulator and capacitors is built, it is likely to be in the wrong place. Thus

FIGURE 10-3 Creating the dimensions for a Eurocard.

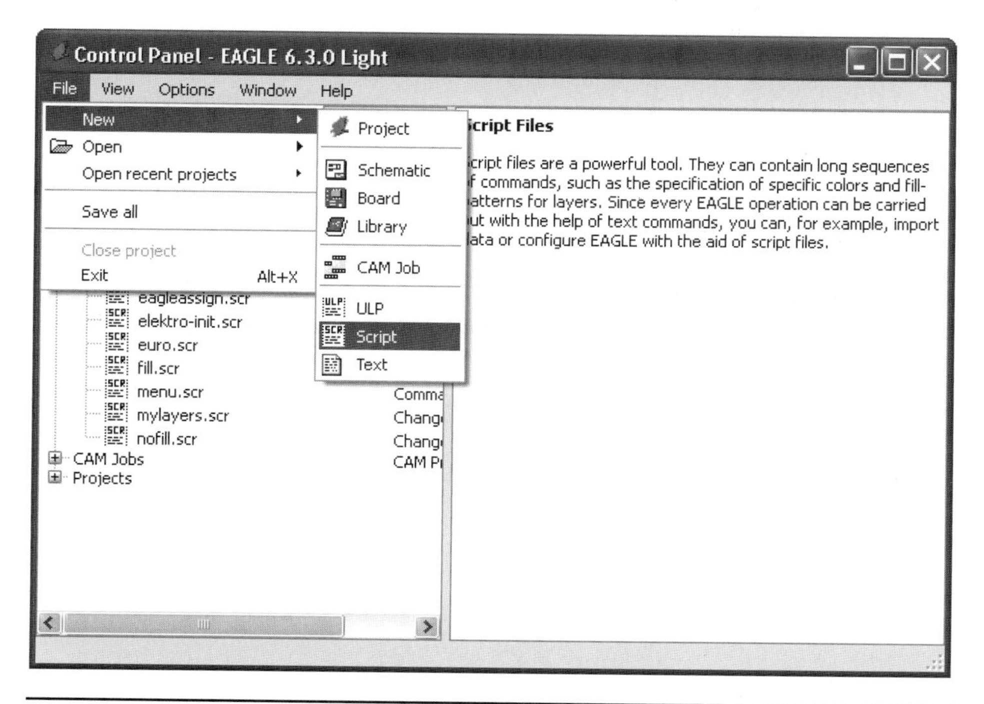

FIGURE 10-4 Creating a new script.

it may as well be built over to the left, out of the way, so that we can move it into place afterward.

The command to create the 7805 is

```
add 7805T@linear (-1 0)
```

You probably will have to explore the library doing a bit of wild-card searching to find the exact location of the component you need. Next, we need to add two capacitors, as recommended in the 7805 datasheet. Make a note of the positions for these capacitors by moving the cursor to where you want them to appear.

```
add C-US050-030X075@rcl (-1.6 -0.3)
add C-US050-030X075@rcl (-0.4 -0.3)
```

The values of these capacitors can be set using the following commands:

```
value c1 330n
value c2 100n
```

To add in the nets, we again have to find the coordinates of the end points and then use them in the following commands:

```
net (-1.4 0) (-1.6 -0.2)
net (-0.6 0) (-0.4 -0.2)
```

The module could do with symbols for GND and 5 V and then connect all these together with nets. Here are the commands for this:

```
add GND@supply1 (-0.2 -0.6)
add VCC@supply1 (-0.2 0.4)
net (-1.6 -0.5) (-1 -0.3)
net (-1 -0.5) (-0.4 -0.5)
net (-0.4 -0.5) (-0.2 -0.5)
net (-0.4 0) (-0.2 0.3)
```

When it is all complete, the area of the schematic should look like Figure 10-5.

Save the file in the `.scr` folder, and then you can use it by clicking the "SCR" button and selecting the script. You can download the script from the book's webpage at www.simonmonk.org.

Scripts are great for simple automation tasks, but they are not able to do looping, branching, or other programming tasks. They also do not allow you to find out things about the schematic before performing some commands relating to the schematic. For example, you cannot select all the resistors and then change their values. To do these more advanced operations, you need to use ULPs.

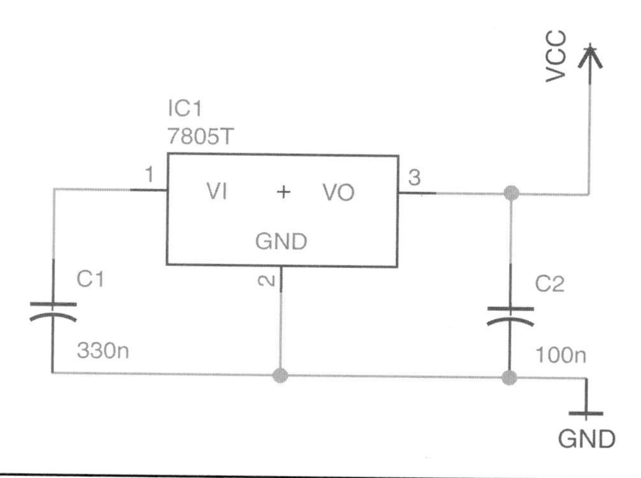

FIGURE 10-5 A 7805 regulator assembly generated by script.

User-Language Programs

The EAGLE ULP is a fully featured programming language with a C-like syntax. If you are used to programming, then you will soon get the hang of the syntax. It is beyond the scope of this book to teach programming in EAGLE ULP, but we will touch on the basics so that you can use the built-in ULPs and also understand a few basic techniques.

Running ULPs

EAGLE has a collection of ready-to-use ULPs. These are accessed from a toolbar icon just next to the "SCR" button. This is labeled "ULP." When you click it, you will be presented with a dialog from which to select the ULP that you want to run (Figure 10-6).

You can also access the ULPs from the EAGLE Control Panel (Figure 10-7), and if you are not sure which ULP you want to run, then this is a better place to do this because together with each ULP is a description of exactly what it does.

These ULPs serve a wide variety of purposes. The ULP selected in Figure 10-8 will generate a bill of materials (BOM) from all the components in a schematic. It will create a separate file detailing all the parts you have used. To run it, right-click on the ULP name in the list, and select the option "Run in Schematic." The window shown in Figure 10-8 will appear, listing all the parts used and giving you the option to save them in various different formats.

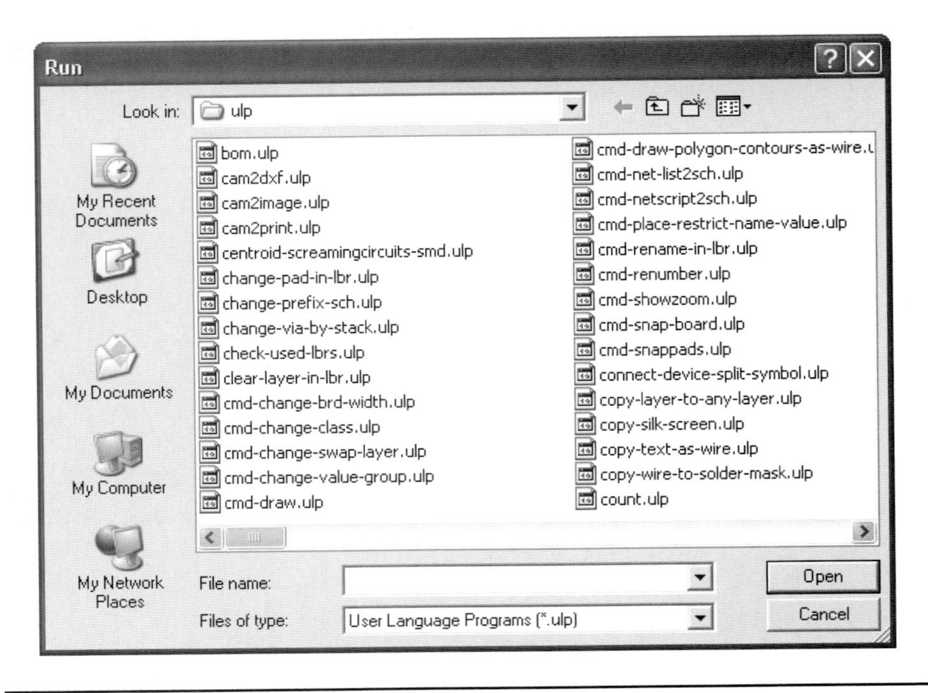

FIGURE 10-6 Running a ULP from the toolbar.

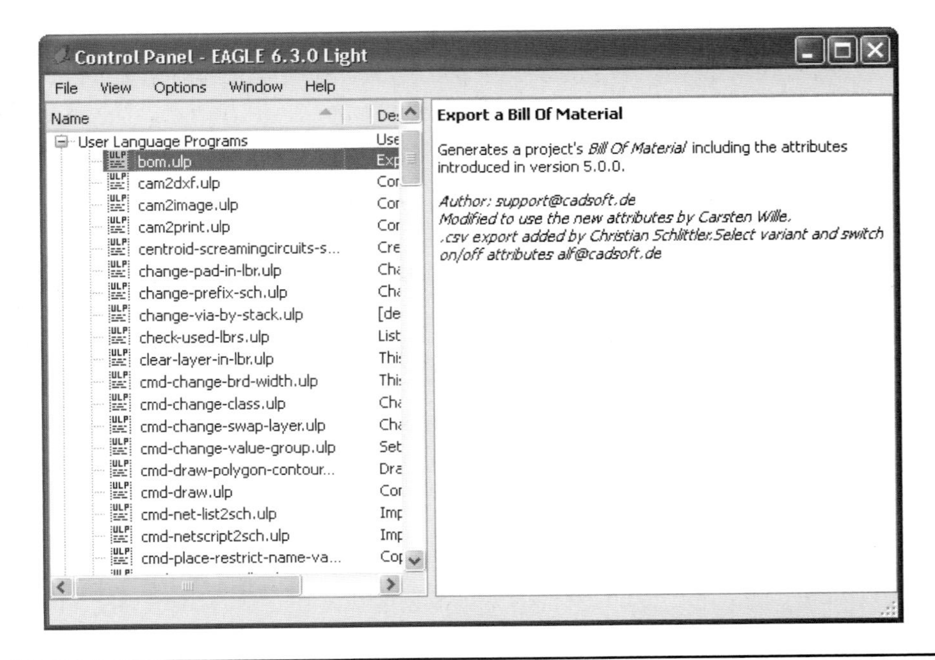

FIGURE 10-7 Running a ULP from the Control Panel.

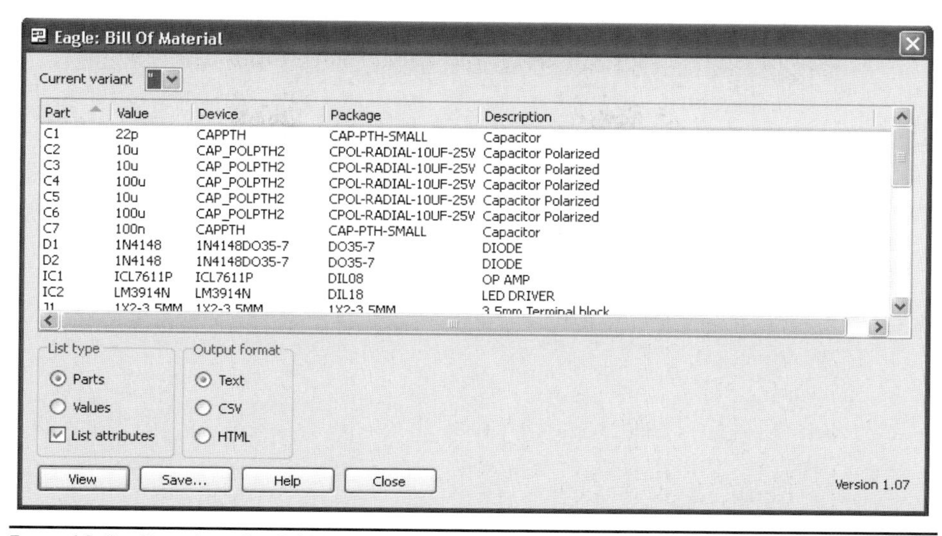

FIGURE 10-8 Running the BOM script.

Clearly, this is a long way from what is possible using simple scripts. Effectively, you can add your own features to EAGLE using ULPs.

The ULP Language

Let's take a look at one of the simpler ULPs (`smash-all-sch.ulp`). This useful ULP applies the Smash command to every part on a schematic. The code for this ULP is

```
#usage "<b>Smash all gates of a schematic</b>\n"
       "<p>"
       "<author>Author: support@cadsoft.de</author>"

// THIS PROGRAM IS PROVIDED AS IS AND WITHOUT WARRANTY OF ANY
// KIND, EXPRESSED OR IMPLIED

string cmd;
string c;

void visible(UL_SCHEMATIC S) {
  sprintf(c, "DISP NONE;\nDISP ");
  cmd += c;
  S.layers(L) {
```

```
         if (L.visible) {
            sprintf(c, "%d ", L.number);
            cmd += c;
            }
         }
      cmd += ";\n";
      return;
    }

    if (schematic) {
     schematic(S) {
       cmd = "DISPLAY NONE 94;\n";
       cmd += "GRID MIL 1;\n";
       S.sheets(SH) {
          sprintf(c, "EDIT .s%d;\n", SH.number);
          cmd += c;
          SH.parts(P) {
             if (P.device.package) {   // **** only Devices with
                                       // Package
                                       // **** without Supply symbol
                                       // Frames ect...
                P.instances(I) {
                   if (I.sheet != 0) {
                      sprintf(c, "SMASH (%.2f %.2f);\n", u2mil(I.x),
                              u2mil(I.y) );
                      cmd += c;
                   }
                }
             }
          }
       }

      sprintf(c, "GRID INCH 0.1;\n");
      cmd += c;
      sprintf(c, "EDIT .S1;\n");
      cmd += c;
      visible(S);
      exit (cmd);
     }
```

```
}

else {
   dlgMessageBox("\n    Start this ULP in a Schematic    \n");
   exit (0);
}
```

Starting at the top of the file, there is a # usage marker. The remainder of that line and the two lines that follow it specify the text that will appear in the Control Panel when you browse for ULPs. As well as plain text, this can also contain HTML; hence the <b> tag will make the first line bold.

Next, two strings are defined. The string `cmd` will eventually contain a whole script full of commands. These are not actually actioned by EAGLE until the following line of code near the end of the file is run:

```
exit (cmd);
```

In other words, the whole ULP builds up a big string of text that is then included in the Exit command that ends the execution of the script and also invokes the command string supplied.

The string `c` is used as a temporary working string whose contents will be built up and then appended to the main `cmd` string.

After the string declarations, there is a definition for a function (`visible`). This function will make the schematic visible and display all the relevant layers. It is called just before the script exits.

This ULP is designed to work on a schematic, and the ULP includes a test to make sure that the window from which the ULP has been invoked is actually a schematic. You will find this in the following line, halfway down the program:

```
if (schematic) {
```

If you mentally close off the braces to find the `else` condition for this, you will find the following lines right at the end of the file:

```
else {
   dlgMessageBox("\n    Start this ULP in a Schematic    \n");
   exit (0);
}
```

This displays a dialog window telling you that you are not in a schematic window.

Assuming that the ULP is being run in a schematic window, we now have a chunk of code that iterates over the schematic in a hierarchical way constructing the command to be run. This uses a syntax that names an object in the design's internal model and then a variable that can be used to access all instances of such objects. For example,

```
schematic(S){
  // do things to every schematic using the variable S
}
```

All the code inside the curly braces will be repeated for each schematic, and the schematic whose turn it is can itself be accessed using the variable S. There is only one schematic, so this syntax makes more sense if we look a bit further down the program, where every sheet of a schematic is iterated over and then within that every part on every sheet using

```
SH.parts(P) {
}
```

Summary

In this chapter we have looked at using scripts and ULPs to automate activities in EAGLE. The EAGLE ULP manual is some 130 pages long, so we have only really scratched the surface of the language. You can find out more about it at ftp://ftp.cadsoft.de/eagle/userfiles/doc/ulp_en.pdf. You will also find a reference for ULP in Appendix C.

In Chapter 11, we will return to libraries and learn how to modify and create our own libraries and parts.

CHAPTER 11
Creating Libraries and Parts

Experienced EAGLE users often find it useful to start building their own library of components that they use frequently. This chapter shows you how to do this and how to copy and modify components from other libraries.

It is easier to copy and modify an existing component than to create a new one entirely from scratch, so we will develop an example where we modify the pad positioning of a part.

Creating a Library

Parts libraries are each held as a single file with the extension `.lbr` in the `lbr` folder within the EAGLE installation directory. New libraries can be added simply by copying `.lbr` files into this directory. In fact, we did this back in Chapter 1 when we installed the Adafruit and Sparkfun libraries. Component manufacturers and suppliers often have their own libraries of components. Another similar library is available from Seeedstudio (www.seeedstudio.com).

However, in this case, you are going to create your own library from scratch. We will give it a general name, and if you do this, you probably will find it to be a useful place to keep modified or new parts that you create.

To create a new library, open the Control Panel window of EAGLE, and then from the File menu, select the option to create a new library (Figure 11-1).

This will open a new untitled Library Editor window. The first thing you should do is save this using the Save command from the File menu with the Library Editor as the current window. You will be prompted for the name of a file. Call it `MyParts.lbr`.

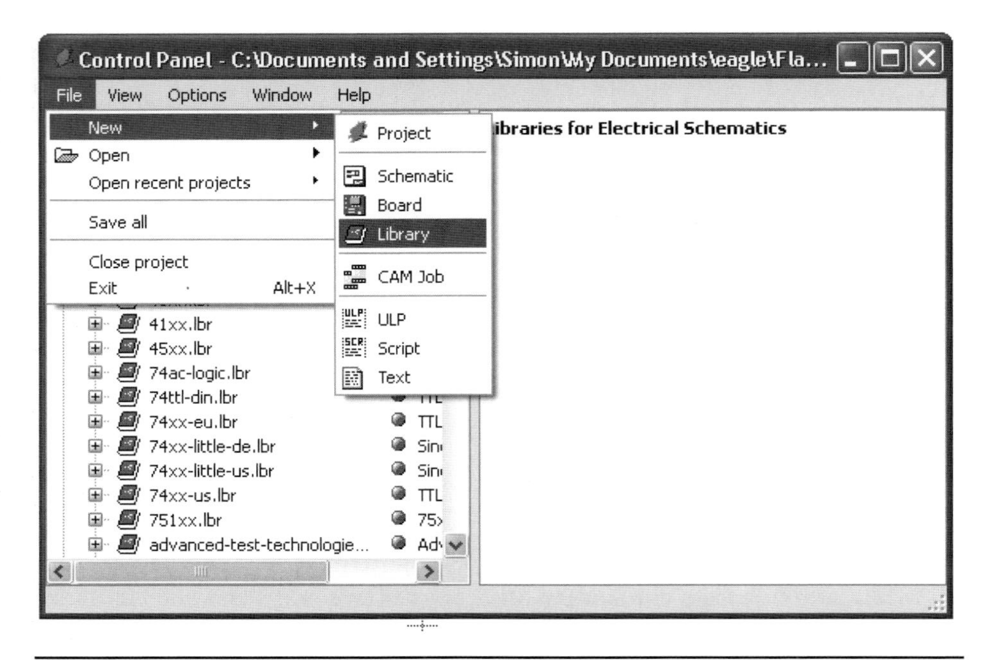

Figure 11-1 Creating a new library.

Copying a Device from Another Library

Later we will look at building a device (part) from scratch, but to start with, we will copy an existing part from another library into our new library. Later we will edit this device.

Copying a part is not quite as simple as selecting a part, doing Copy, moving to the library where you want to have a copy of it, and doing Paste. Instead, EAGLE has the concept of a *current* library. Because we have just created a new library, this will be the current library. To copy the part into the current library ("MyParts"), we first need to go and find the part that we want to copy in the "Libraries" section of the Control Panel.

The first part we are going to copy is the LED display module that we used back in Chapter 8. You may remember that this had some mislabeled pins. You will find this part in the Sparkfun "Displays" library, where it is called `7-SEGMENT-4DIGIT-YOUNGSUN`. Right-click on it, and select the option "Copy to Library" (Figure 11-2).

As soon as you do this, the part will appear in the Part Editor window (Figure 11-3).

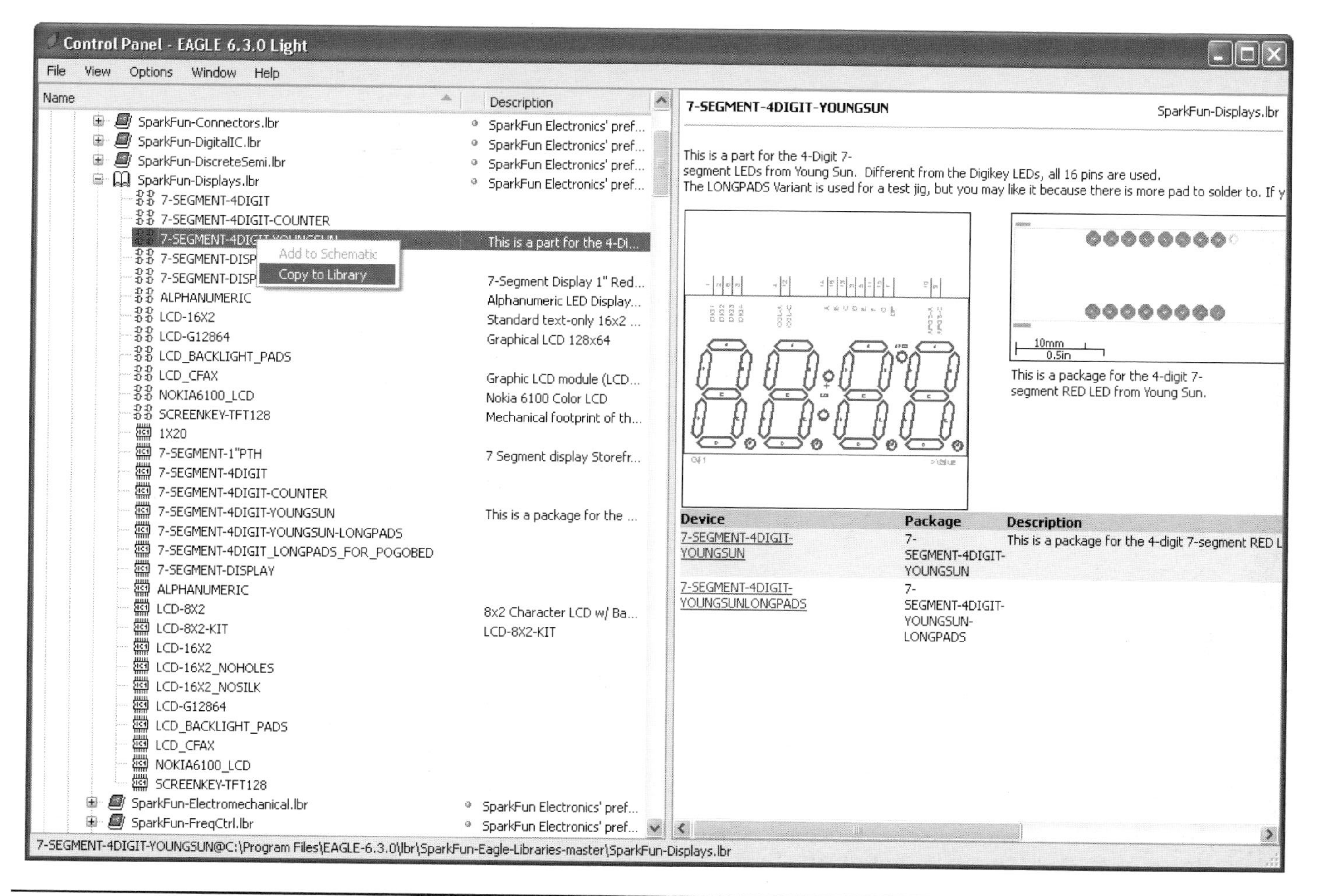

FIGURE 11-2 Copying a part.

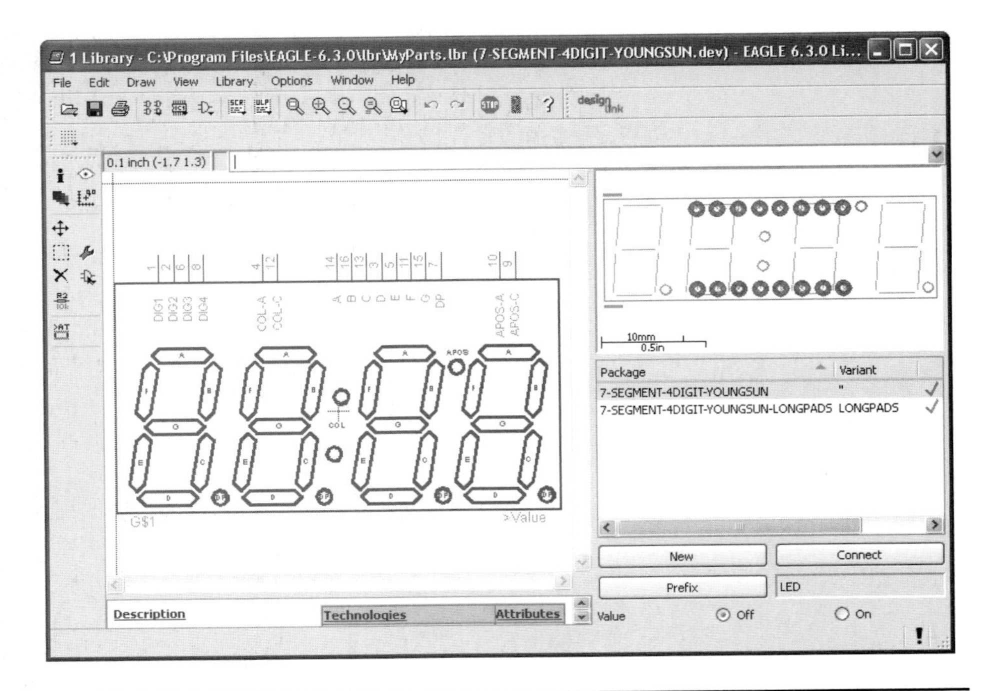

FIGURE 11-3 The Part Editor window.

The Part Editor

Let's take a quick tour of the Part Editor window. The large editor area shows the schematic view of the part (its symbol). To the bottom right, there is a list of packages available for this symbol. This is showing two entries, one normal and one with long pads. Above this is the package as it would appear on the board layout.

You will notice that over on the left of the editor we have a Command toolbar that operates on the same principal as the Schematic and Board Editors but with a smaller set of commands available.

Devices, Symbols, and Packages

Before we start modifying this part, we need to understand a bit more about how parts are organized in an EAGLE library. A *device* has a symbol (this is shown in the schematic); it also has one or more *packages*. These will appear on the board layout. The part as a whole is represented as a device. The device links together the symbol and the packages, as well as specifying the relationship between the pins on a symbol and the pads on a package.

Symbols and packages are created independently and linked together by devices, allowing the same package design to be used by a number of different symbols.

In Figure 11-3, the Part Editor is showing the device. This same editor window is also used when you want to edit the symbol or one of the packages of the device. You will notice three new icons on the toolbar to the right of the "Printer" icon that allow you to select the element of the device that you want to edit.

Editing a Part

We need to fix the labels on this part because the anode and cathode of the colon and apostrophe of the display are swapped over. You may be wondering if this is a change to the pad names or to the symbol. Well, the pad names are just named 1 to 16 on both packages, so the change needs to be made in the linkage between the pin names on the symbol and the pad names. This takes place for each package of the device. So let's start with the package `7-SEGMENT-4DIGIT-YOUNGSUN`. Select this in the list of packages, and then click the "Connect" button. This will open the window shown in Figure 11-4.

The rightmost of the three columns shows the current linkages between pins and pads for that package. In this case, you can see that APOS-A is linked to pad 10 and APOS-C is linked to pad 9. These need swapping over, as do the associations between COL-A (pad 4) and COL-C (pad 12). To sort this out, we first need to disconnect these four connections. To do this, select each in turn, and click "Disconnect." This will have the effect of moving both the pins and the pads concerned into the first two columns, as shown in Figure 11-5.

We now need to remake these links, but this time linking

- APOS-A to pad 9
- APOS-C to pad 10
- COL-A to pad 12
- COL-C to pad 4

This is done by selecting one pin from the first column and one pad from the second column and then clicking "Connect." So click APOS-A and pad 9 and click "Connect." Repeat for the other three links above. When everything is connected,

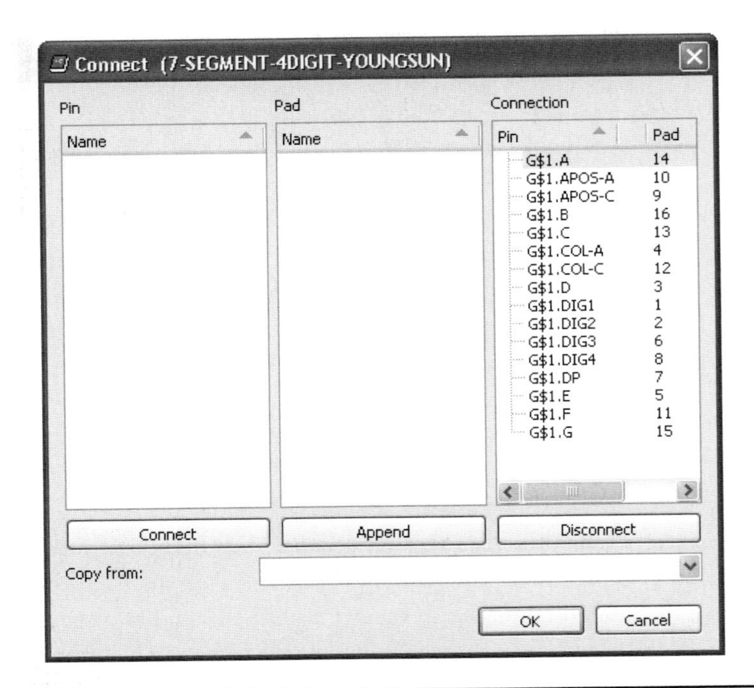

FIGURE 11-4 Editing connections.

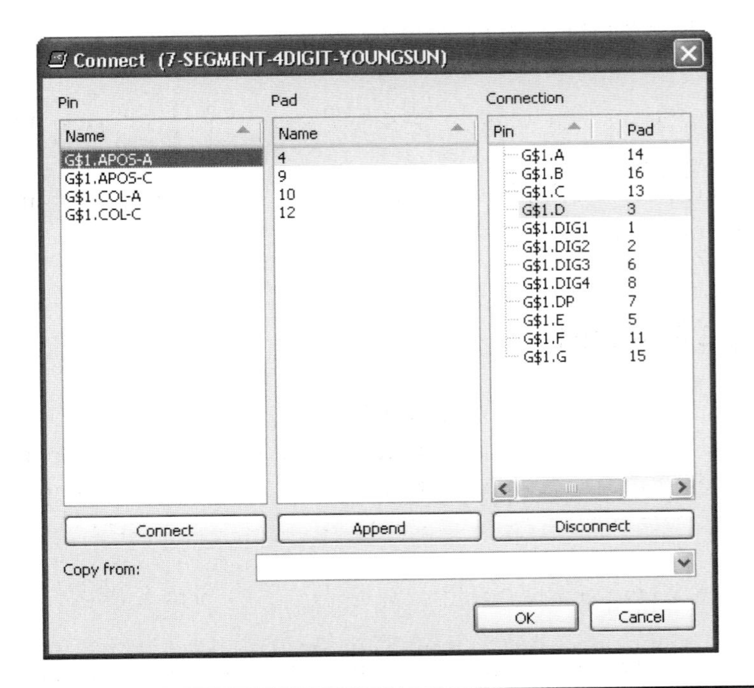

FIGURE 11-5 Rearranging pads and pins.

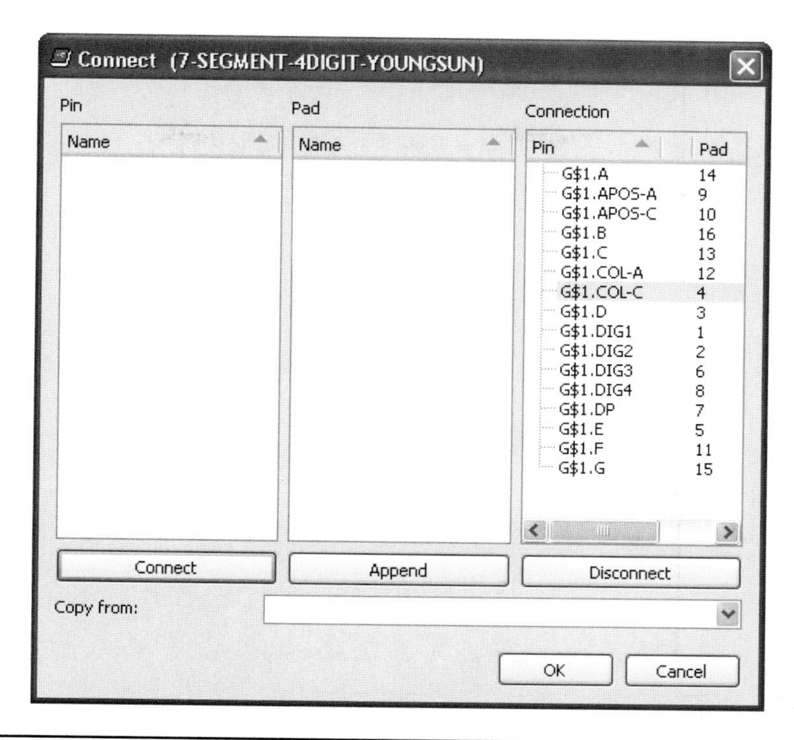

FIGURE 1-6 The connections amended.

the Connection editor should look like Figure 11-6, with everything back in the right-hand column.

Finally, repeat this for the other package in the device, and then save the device from the File menu.

Creating a New Part

To really get to grips with the Part Editor, we are now going to add a completely new device to our library. The device we are going to add is shown in Figure 11-7.

This device is a vacuum tube (a double triode). The vacuum tube is a device largely made redundant by transistors and integrated circuits (ICs) but retaining a niche role in audio amplifier design, where it is used by audiophiles for the low-frequency even harmonic distortion that many people find gives an attractive color to music.

It is not a total surprise that this device is not currently included in the libraries that ship with EAGLE, so we will make our own part for it, especially because I

FIGURE 11-7 A 12AU7 triode tube.

intend to make myself a preamp using this device and would like to design the PCB using EAGLE.

You can find the datasheet for this tube at www.wooaudio.com/docs/tube_data/12AU7.pdf.

The steps involved in making this part will be

1. Make the package.
2. Make the symbol.
3. Make the device to link the two together.

Making a Package

If it is not already open, then open your "MyParts" library. To create a new package, you need to click on the "Packages" icon on the toolbar. This will open a window that shows the packages that we copied for the LED display (Figure 11-8). In the field labeled "New," enter the name `RETMA-9A`. This is the name given to the package on the tube's datasheet. Hit the "Enter" key, and the Part Editor window will open on a new file called `RETMA-9A.pac`. We will use this editor to create the package.

The package footprint consists of nine pads arranged around a circle with a diameter of 0.5 in. The pads are spaced out as if there are 10 pins but with one of the pads missing. A good technique is therefore to draw a circle (that we can later delete) of diameter 0.5 in. and then position the pads around it.

It does not really matter what layer we draw the circle on because we are going to delete it later, but in this case, it is on "tPlace." Because there are effectively 10 pin positions, each pin will be 360/10 = 36 degrees from the next pin around the center of the package. If you use the Mark tool and mark the origin of the package, then as well as showing the *X* and *Y* coordinates of the cursor relative to the center of the circle, it will also show the polar coordinates, that is, the distance from the center and the angle. We can use this to position the pads at 0, 36, 72,

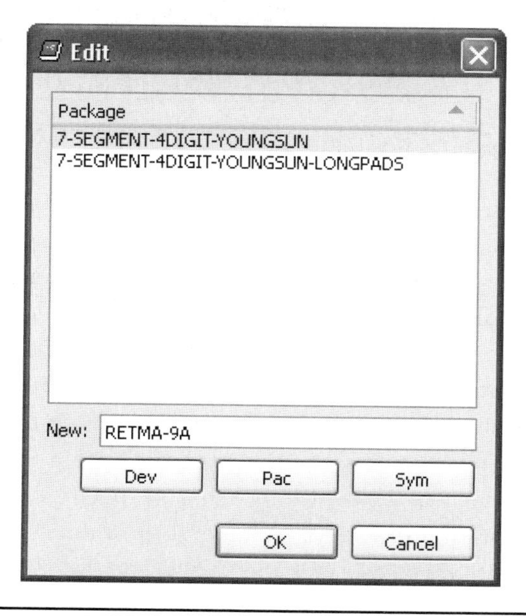

FIGURE 11-8 Creating a new package.

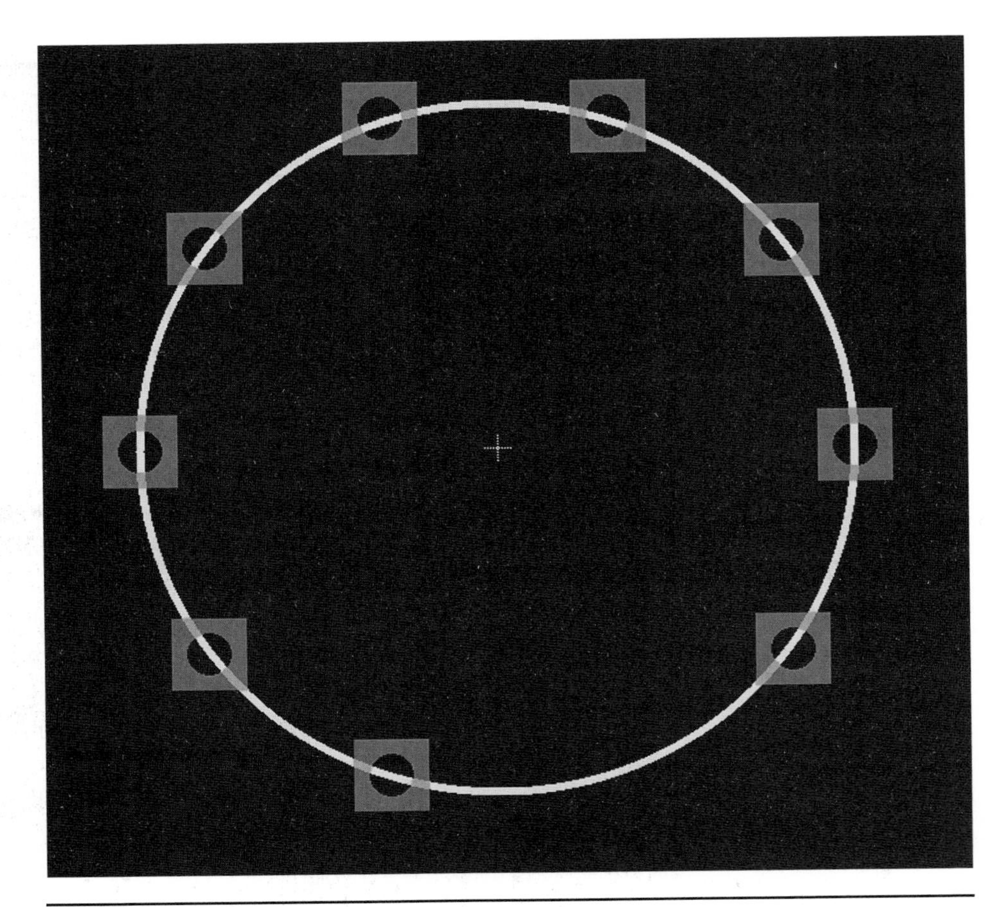

Figure 11-9 Positioning pads.

and so on degrees around the circle (Figure 11-9). To do this, use the Pad command from the Command toolbar. You can change the diameter and drill to match the tube socket that you are going to use before adding the pads. You will need to set the grid to be, say, 0.001 in. in order to position the pads accurately.

The tube itself has a diameter of about 0.75 in., so let's mark that footprint by drawing a circle with that diameter in place of the 0.5-in. circle, which we can now delete. Note that you probably will want to reset the grid to 0.025 in.

The next step is to name all the pads. We are just going to name these sequentially from 1 to 9 clockwise around the package. We now need to add two pieces of text onto the package ">Name" and "Value." The text ">Name" should be placed on the "tNames" layer and "Value" on the "tValues" layer. Once all this is complete, the package should look something like Figure 11-10. Don't forget to save it.

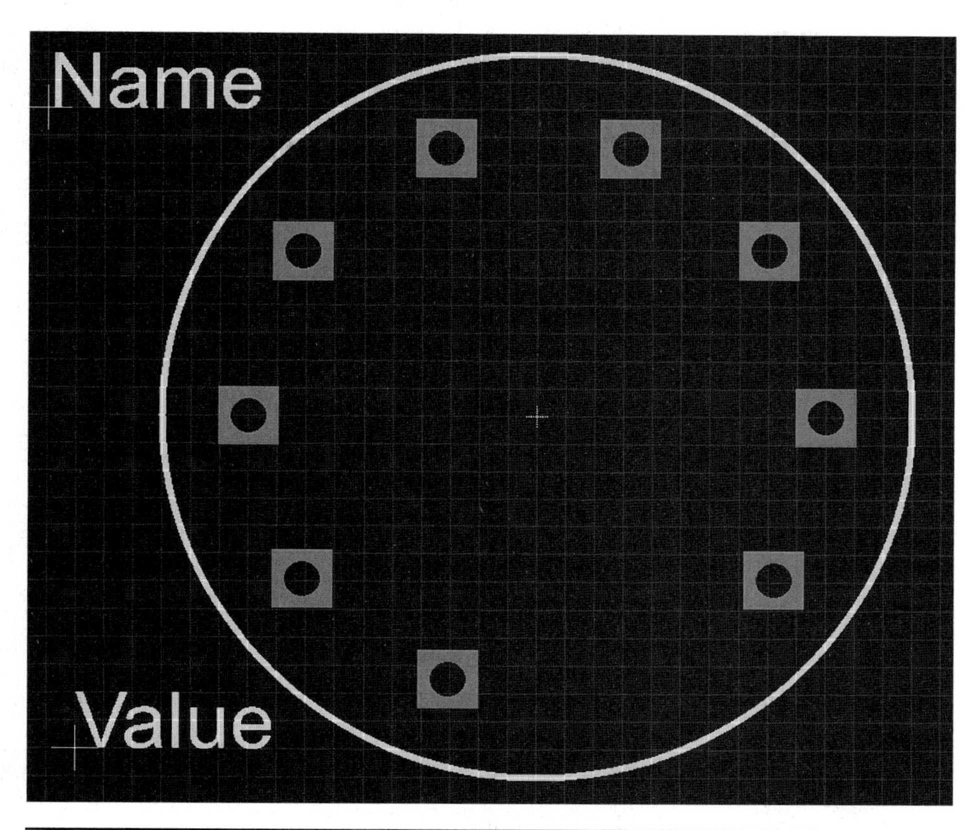

FIGURE 11-10 The completed package.

Making a Symbol

To make the symbol for the device, we are going to carry out a similar process to making the package. However, one key difference is that this tube is a dual triode; that is, it is a single glass envelope containing two triodes. We therefore only need to create the symbol for one triode, and we will see later how these can be added twice to the device as *gates*. But first, let's draw the symbol.

Create a new symbol called `TRIODE` by clicking on the "Symbol" icon and then typing the name `TRIODE` in the "New" field and hitting the "Enter" key. This will display a blank editor window for us to use on the file `TRIODE.sym`. The circuit symbol for a triode is a circle containing what are in essence three horizontal lines from top to bottom, the plate, grid, and cathode. The grid is a dashed line, and the cathode has turned-over corners. In any case, it is pretty easy to draw.

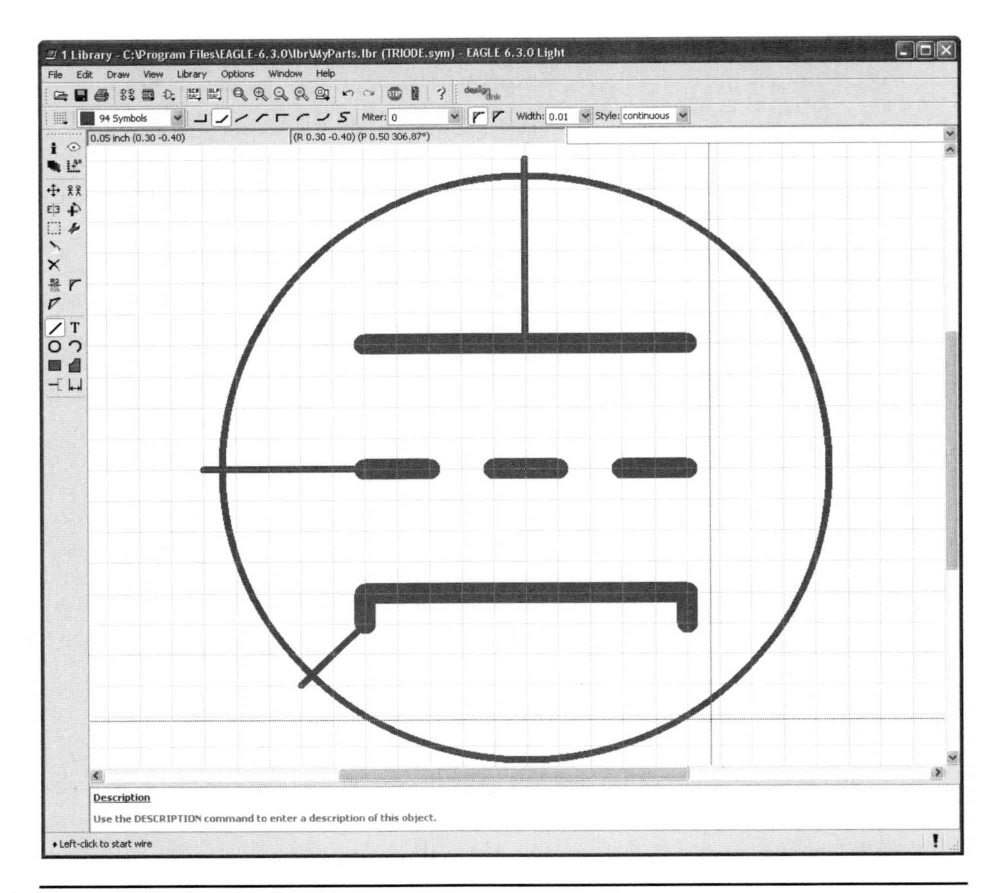

Figure 11-11 Drawing the triode symbol.

The size is pretty arbitrary, but I am going to make the symbol about ¾ in. wide because these will be the central components of the amplifier that I intend to build. Draw the symbol on the "94 Symbols" layer using the Circle and Wire commands. I used a wire width of 0.32 in. There also need to be wires of 0.01 in. from each of the three lines out to the edge of the circle. Figure 11-11 shows the remainder of the "Symbol" layer drawing.

Just as we had to add pads to the package, we now have to add pins to the symbol. Select the Pin command, and then select a style of pin from the toolbar that does not include a line attached to the pin. That is, select a style that is a simple circle ("Point"). The pins will need renaming and their direction type setting. The most convenient way of changing this for each pin is to use the Information tool.

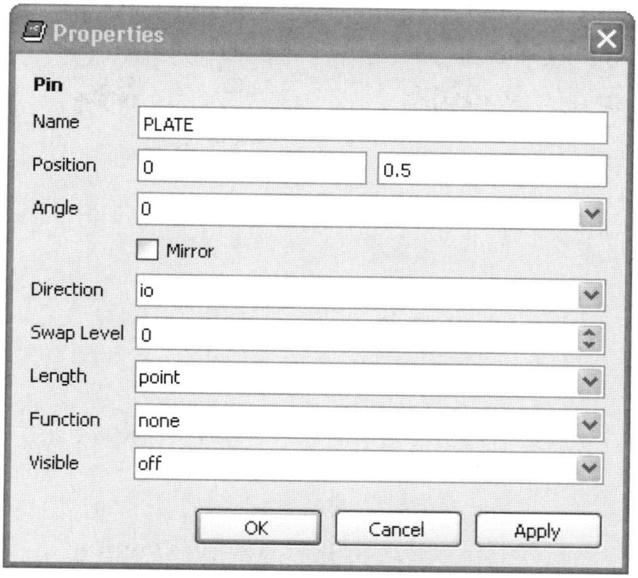

Figure 11-12 Changing a pin.

Starting with the pin you added at the top of the symbol, select the Information command, and then click on the pin. Then change the name of the pin to `PLATE`, the direction to `pass`, and "Visible" to `off`, as shown in Figure 11-12.

The various directions allowed for a pin are used by the electrical result checker (ERC). Thus, for example, it will flag outputs being connected together on a logic gate. In this case, we are specifying that the pin is electrically speaking passive (`pass`). This isn't really true, but selecting this option will prevent any false warnings from the ERC. Repeat this process for the GRID and CATHODE pins.

You will have noticed an area at the bottom of the parts window suggesting that you `Use the DESCRIPTION command to enter a description of this object`. To do this, click on the "Description" hyperlink immediately above that message. This will enter a new window into which you can enter a description of the symbol.

In addition to the two triodes, the tube also has three connections used to provide power to the heater element, one of the connections being a center tap. We therefore need to bring the three "power" pins that we can collect together into a third symbol. Even though there are two triodes in the glass tube, they have shared connections for the heater power.

Create a new symbol, and call it `HEATER_SUPPLY`. Place three pins on the design, and change their names to "HA," "HC," and "HB" (heater A, B, and

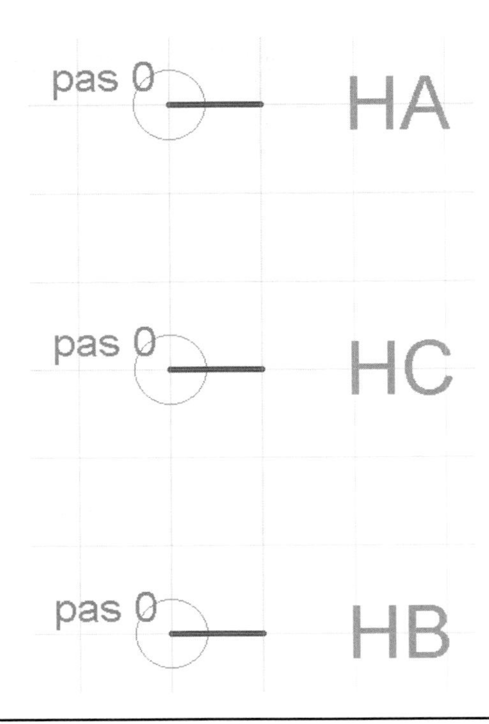

Figure 11-13 Creating a heater supply symbol.

common). Pick a pin style with a bit of a tail on it. The end result of this is shown in Figure 11-13.

Making a Device

We have a package and a symbol for our valve, but we have not linked the two or let EAGLE know that the package actually contains two triode symbols. To do this, we need to create a device by clicking on the "Devices" icon in the toolbar and then entering the name `12AU7` into the "New" text field. In the view that then appears, we need to use the Add tool to add two of the `TRIODE` symbols and one `HEATER_SUPPLY` symbol to the main editor area.

Next, we need to add the package, so, in the bottom right of the Editor Window, click on the "New" button, and select "RETMA-9A." The editor window should now look like Figure 11-14.

There is still no link between the pins on the symbols and the pads on the package, so let's make those connections by clicking on the "Connect" button. After matching the pairs up using the datasheet, it will look like Figure 11-15.

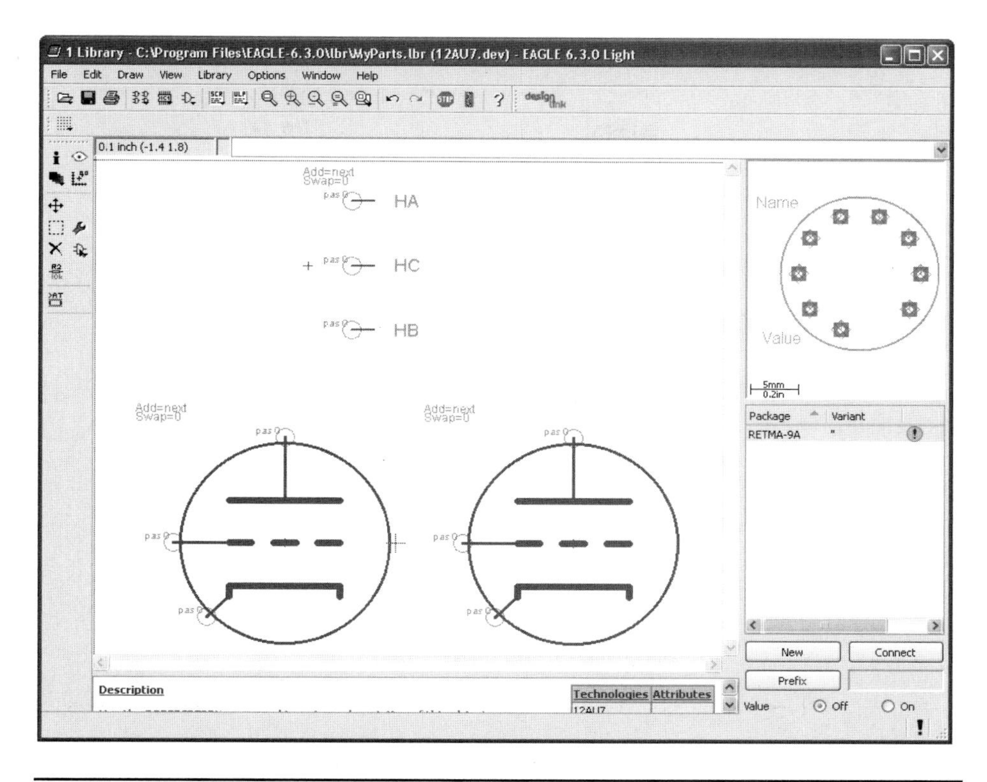

FIGURE 11-14 The device with symbols and package.

Our part is now complete. Don't forget to save it. You can try adding it to a schematic. Remember that because the device has a number of gates, the first two times that you add it to the schematic, it will add a triode symbol, and the third time it will add the heater supply symbol. You can also switch over to make sure that the board layout looks okay too. Figure 11-16 shows the device as used in a schematic.

If you want to share the library containing the part, you can simply pass on the "MyParts" library file that you will find in the `lbr` folder in your EAGLE installation directory. Note that if you are planning on sharing the part, you might want to give it a more specific name.

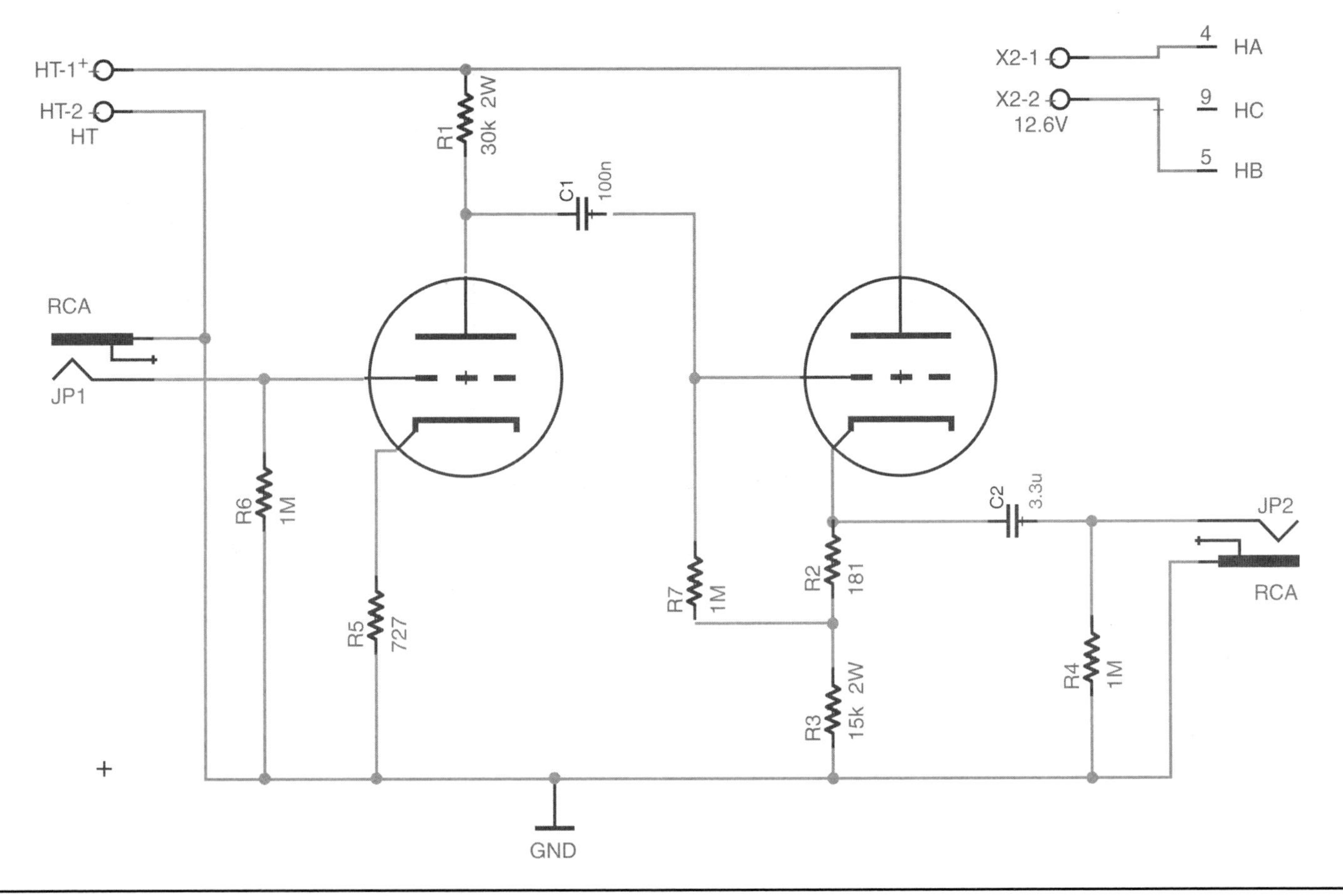

FIGURE 11-15 Making connections from pins to pads.

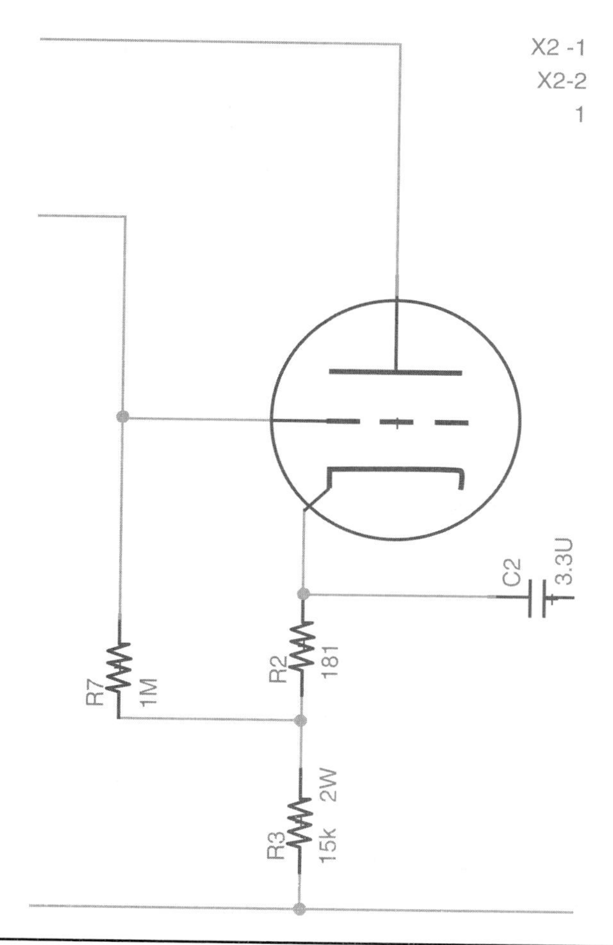

Figure 11-16 Using the new device.

Summary

In this chapter, you have discovered how to create your own library, to which you can either add modified packages, symbols, and devices or create your own parts from scratch. Before creating a new part, it is always worth an Internet search to see if someone else has already made that part. Some component suppliers also provide an EAGLE library for their products.

APPENDIX A
Resources

EAGLE is a complex and powerful piece of software, and a book of this size cannot explore every feature of this tool. In this Appendix, you will find links to more resources that will help you to use and learn more about this software.

Official Documentation

The official manual for EAGLE is available as a free download from www.cadsoftusa.com/training/manuals. This is a large and comprehensive manual that documents everything about EAGLE in a very methodical manner. It is very much a reference book rather than a "how to" book, so once you have mastered the basics, it will help you to find chapter and verse on a particular feature of the software.

Forums

The largest and official forum for EAGLE can be found at www.element14.com/community/community/knode/cadsoft_eagle/forums/. It is very active and a useful resource. If you have a problem, try searching, and you may well turn up an answer. If no answer is forthcoming, then post something—people are happy to help.

Tutorials

This book has a tutorial and practical style, but sometimes seeing someone actually do something can make things clearer. There are many videos on using EAGLE. You will find lots of these by searching on YouTube.

Jeremy Blum has produced an excellent series of videos on using EAGLE that can be found at www.youtube.com/watch?v=1AXwjZoyNno. Sparkfun also has a series of useful tutorials on using EAGLE that you find at www.sparkfun.com/tutorials/115.

Sources of Library Parts

The official site is also a great place to go looking for libraries (www.cadsoftusa.com/downloads/libraries). You can also upload your own libraries to this site so that others can use them.

APPENDIX B
EAGLE Layers

The most commonly used layers are described in Chapters 4 and 5. The following table provides a more complete list of the layers.

Layers Used in the Layout Editor

You will find more description on how to use the more common of these layers in Chapter 4.

Layer #	Name	Description
1	Top	The top copper layer
2–15	Route2–Route15	Inner copper layer for boards with more than two copper layers; not used in EAGLE Light edition
16	Bottom	The bottom copper layer
17	Pads	Solder pads
18	Vias	Vias linking one layer to another
19	Unrouted	Air wires showing as yet unrouted connections
20	Dimension	The board outline including holes
21	tPlace	The top silk-screen layer
22	bPlace	The bottom silk-screen layer
23	tOrigins	Automatically generated markers for items on the top layer
24	bOrigins	Automatically generated markers for items on the bottom layer

(continued on next page)

Layer #	Name	Description
25	tNames	Names of components on the top layer; if you want these to appear on the silk screen then include this layer in the CAM job or move the text to the tPlace layer
26	bNames	See above but for the bottom layer
27	tValues	Component values on the top layer; if you want these to appear on the silk screen then include this layer in the CAM job or move the text to the tPlace layer
28	bValues	See above but for the bottom layer
29	tStop	Solder stop mask corresponding to the laquer layer over most of the copper tracks on the top layer except the pads
30	bStop	See above but for the bottom layer
31	tCream	The area to be "tinned" with solder or for solder paste to be applied, depending on the manufacturing process
32	bCream	See above but for the bottom layer
33	tFinish	Used if some areas require a special treatment—say gold plating of edge connectors.
34	bFinish	See above but for the bottom layer
35	tGlue	Used in SMT to indicate location of glue to attach the component to the top side of the baord before cooking
36	bGlue	See above but for the bottom layer
37	tTest	User definable layer intended for test purposes
38	bTest	As above
39	tKeepout	A polygon in this layer prevents placement of components in the area on the top layer; for example, for a heatsink area
40	bKeepout	See above but for the bottom layer
41	tRestrict	A polygon on this area prevents any tracks being routed through the area on the top layer
42	bRestrict	See above but for the bottom layer
43	vRestrict	See above but applied to vias
44	Drills	Used for pads (through hole) and vias
45	Holes	Mounting holes in the PCB not associated with pads or vias
46	Milling	Areas of the board to be cut out; normally you would use the dimension layer for this
47	Measures	User definable layer intended for measurement purposes
48	Document	Documentation layer—annotations to the design

Layer #	Name	Description
49	ReferenceLC	Reference marks
50	ReferenceLS	Reference marks
51	tDocu	Top layer documentation
52	bDocu	See above but for the bottom layer

Layers Used in the Schematic Editor

See also Chapter 4 for more information on using these layers to create schematic diagrams.

Layer #	Name	Description
91	Nets	Nets connecting component pins
92	Buses	Clustered sets of nets; for example, a parallel data bus from one IC to another
93	Pins	Connection points on a component symbol
94	Symbols	Component symbols
95	Names	Component names
96	Values	Component values
97	Info	Additional information about components
98	Guide	Guidelines for symbol alignment and selection

APPENDIX C

User-Language Program Reference

Although the user-language program (ULP) language is C-like, it is not exactly C. For example, the equivalent to the C "float" type is called "real" and there is no "boolean" type. This appendix gives a summary of the ULP language, in particular where it differs from the C syntax.

You will find information on using ULPs in Chapter 10. You will also find the built-in help files of EAGLE (accessible from the Help menu) useful.

Data Types

The following primitive types are defined.

Type	Description	Example
char	A single ASCII character; constants are delimited with single quotes	char ch = 'A'
int	A 32-bit integer	int x = 123
real	A 64-bit float	real x = 123.456
string	A dynamically allocated string of characters (see next section)	string s = "abc"

Strings

The "string" data type is more like a Java "string" class than a C character array. You can concatenate strings using the + operator and also use a number of built-in functions to manipulate strings.

Function	Description	Example
strchr	Returns the position of the first occurrence of a character within a string, or returns –1 if it's not found	string s = "test"; int p = strchar(s, 'e');
strjoin	Concatenates arrays of strings; see strsplit for the reverse process	string a[] = { "aa", "bb", "cc" }; string s = strjoin(a, ',');
strlen	Returns the length of the string	string s = "test"; int p = strlen(s);
strlwr	Returns a copy of the string in lowercase	string s = "TeST"; string s2 = strlwr(s);
strrchr	See strchr but the last occurrence rather than the first	
strstr	Returns the start position of the first occurrence of a substring within a string, or –1 if it's not found; it takes an optional third parameter of the start index	string s = "hello world"; int x = strstr(s, "world");
strsplit	Splits a string into a string array using the separator character specified; it returns the number of array elements created	string s = "aa,bb,cc"; string a[]; strsplit(a, s, ',');
strsub	Returns the substring of a string within bounds specified; the parameters are start position and length of the substring	string s = "Hello World"; string s2 = strsub(s, 3, 2);
strtod	Convert a string to a real	real x = strtod("12.34");
strtol	Convert a string to an int	int x = strto1("12");
strupr	See strlwr but to uppercase	
strxstr	See strstr but the search string can be a regular expression	

Arrays

Arrays use the same [ ] syntax as C, but can be created dynamically from built-in functions, as the example below of splitting a string into a string array illustrates.

```
string a[];

int n = strsplit(a, "Field 1:Field 2:Field 3", ':');
```

Logical and Bitwise Operators

These are the same as C. So, bitwise "and" and "or" are "&" and "|" respectively. Logical equivalents are: "&&" and "||". The not operator is "!" and xor id "^".

Control Structures

These are the same as C. You have access to the usual control structure commands such as "if", "while", "for", "break", etc.

Special Constants

A number of EAGLE specific constants are defined for the ULP.

Constant	Description
EAGLE_VERSION	The version of EAGLE as an int; e.g., 6
EAGLE_RELEASE	The release of the EAGLE version above; e.g.; 3 for EAGLE 6.3
EAGLE_SIGNATURE	Descriptive string of the EAGLE version
REAL_EPSILON	The smallest possible real number that is greater than 0
REAL_MAX	The maximum real value
REAL_MIN	The minimum negative real number
INT_MAX	The maximum int
INT_MIN	The minimum negative int
PI	The numeric constant pi

Dialog Functions

The ULP language includes a number of built-in dialog types that you can use in your programs.

- **dlgDirectory**—Prompts for a directory from the file system and returns the path
- **dlgFileOpen**—Prompts for a file to open and returns the file path
- **dlgMessageBox**—Displays a message to the user

See the online help for the full syntax for these commands.

Other Built-in Functions

There are many other built-in functions available in the ULP language, for tasks such as reading and writing files, performing HTTP requests, and even manipulating XML. See the online documentation for these functions in the built-in help system.

Index

References to figures are in italics.